Supersymmetric Quantum Mechanics

An Introduction

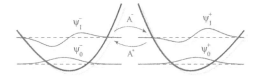

Supersymmetric Quantum Mechanics

An Introduction

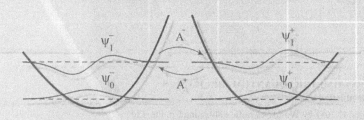

Asim Gangopadhyaya • Jeffry V Mallow
Loyola University Chicago, USA

Constantin Rasinariu
Columbia College Chicago, USA

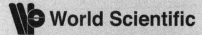 **World Scientific**

NEW JERSEY · LONDON · SINGAPORE · BEIJING · SHANGHAI · HONG KONG · TAIPEI · CHENNAI

Published by

World Scientific Publishing Co. Pte. Ltd.

5 Toh Tuck Link, Singapore 596224

USA office: 27 Warren Street, Suite 401-402, Hackensack, NJ 07601

UK office: 57 Shelton Street, Covent Garden, London WC2H 9HE

British Library Cataloguing-in-Publication Data
A catalogue record for this book is available from the British Library.

SUPERSYMMETRIC QUANTUM MECHANICS
An Introduction

ISBN-13 978-981-4313-08-7
ISBN-10 981-4313-08-4
ISBN-13 978-981-4313-09-4 (pbk)
ISBN-10 981-4313-09-2 (pbk)

Printed in Singapore.

DEDICATION

AG: To parents: Shefalika & Anil Ganguly and Madhuri & Phanindra N. Dasgupta.

JVM: To my wife and son, Ann and David Mallow, and to the memory of my sister, Brona Neville-Neil.

CR: To my wife, Mariana.

Preface

This book is an outgrowth of a seminar course taught to physics and mathematics juniors and seniors at Loyola University Chicago. Our goal is to provide a single compact source for undergraduate and graduate students and professional physicists who want to understand the essentials of supersymmetric quantum mechanics (SUSYQM) and its connections to quantum mechanics itself.

The book covers a wide range of topics. We have striven to make it as flexible as possible. The material is divided into four sections: fundamentals, applications, summary, and solutions to the problems.

In the first section we review those parts of quantum mechanics related to SUSYQM, as well as the basics of SUSYQM. For example, in Chapter 2 we investigate the infinite square well, followed by the much less well-known $cosec^2$ potential, to demonstrate the startling near-identity of their eigenvalues. We also investigate the harmonic oscillator, the "proto-supersymmetric" system, as it might be designated. We then consider the hydrogen atom, not simply for its own sake, but because it presages our study of Dirac theory and its connection to SUSYQM in the applications section. We have modified the traditional order of some topics to maintain our goal of connecting ordinary quantum mechanics to SUSYQM. For example we first show that different potentials have almost identical spectra, and then, in a subsequent chapter we discuss the more general concepts regarding measurements, expectation values, or mixed states. The reason is that we want to emphasize the essential hamiltonian degeneracy that is fundamental in SUSYQM.

The second section is dedicated to selected applications such as generating orthogonal polynomials by shape invariance, the supersymmetric version of the WKB approximation (SWKB), scattering, isospectral deformations, and the supersymmetric quantum Hamilton-Jacobi formalism. This section is relevant not only to the advanced student who is interested in gaining insight into the current SUSYQM research, but also to the professional physicist who is looking for a good reference source.

Each of the chapters contains problems. All are designed to give the reader hands-on experience with SUSYQM. We conclude the book with a summary and problem solutions.

The references at the end of the book mostly represent the sources we have consulted in writing it. They are collected together to help the reader orient herself/himself to the vast literature that is available. The list is by no means exhaustive, and we apologize for the many omissions.

Acknowledgments

AG: I would like to acknowledge support from many that made this project possible. My parents who, through their massive sacrifices, instilled in me the value of learning. My mentors, Quraishi Guruji, Ratna K. Thakur, Carl M. Shakin and Uday P. Sukhatme who helped shape my academic life. My children - for their love and their irreverence that has helped me keep a clearer perspective. Finally, my ultimate teacher, my wife Alpana. Without her support and encouragement none of this would be possible.

JVM: My wife Ann is a constant source of inspiration and support. In addition, her copy-editing of the book for language and style has greatly improved it, and her curiosity about its contents has put me through my pedagogic paces. For all of this, I am grateful.

CR: I would like to thank my mentors Miron Bur, Dan Albu, Bernhard Rothenstein, Henrik Aratyn, and Uday P. Sukhatme for their guidance. My parents Traian and Paraschiva Răşinariu for their constant encouragement to learn, and my children and wife, for their infinite love and patience.

Contents

PART 1

Fundamentals of
Supersymmetric Quantum Mechanics

Chapter 1

Introduction

Since the advent of quantum mechanics with the Schrödinger equation in 1926, and its relativistic counterpart, the Dirac equation two years later, physicists have been intrigued by some of quantum mechanics' odd properties. These include:

- Accidental degeneracies. Energies often turn out to be strangely independent of some quantum numbers, or dependent on unexpected combinations of quantum numbers.
- Energy coincidences. Very different potentials give rise to the same spectrum of allowed energies.
- Suggestive factorizations. The hamiltonians for certain problems look as if they can be split into two factors, but with a critical "leftover."

Dirac is usually credited with introducing the factorization method for the harmonic oscillator in 1928; he, however, credits Fock with that discovery [1]. Schrödinger himself noticed some of the symmetry features of solutions to his equation, and provided a way to factorize the hamiltonian for the hydrogen atom and other potentials in 1941 [2]. A decade later, Infeld and Hull generalized this to numerous other systems [3] (a set of systems now known as shape invariant potentials.) All of these observations turned out to be hidden manifestations of an underlying symmetry.

---

[1] P. A. M. Dirac, Communications of the Dublin Institute of Advanced Study, **A1**, 5-7 (1943). Dirac references Fock, *Zeits. f. Phys.* **49** 339 (1928).

[2] E. Schrödinger, " Further studies on solving eigenvalue problems," *Proc. Roy, Irish Acad.*, **46A**, 183-206 (1941).

[3] L. Infeld and T.E. Hull, "The factorization method," *Rev. Mod. Phys.*, **23**, 21-68 (1951).

The next impetus came in the early 1980's, when elementary particle physicists attempting to find an underlying structure for the basic forces of nature, proposed the existence of "shadow" partner particles to the various known (or conjectured) elementary particles. Thus, to the photon is conjectured the photino; to the quark-binding gluon, the gluino; to the as-yet-unobserved graviton, the gravitino; and to the exotic W particle, the wino (pronounced *wee-no*). These particle partnerships comprise what is called supersymmetry.

Supersymmetry with its new mathematical apparatus soon led to investigations of partnerships in other areas of physics. One was traditional quantum mechanics itself, where, as we have noted, a variety of mysterious coincidences had long been known. First employed as a so-called "toy model" of field theory, *supersymmetric quantum mechanics*, based on the notion of "partner potentials" derivable from an underlying "superpotential" was born. Supersymmetric quantum mechanics was soon found to have value in its own right, with application to the resolution of numerous questions in traditional quantum mechanics and the posing of various interesting new ones.

In this chapter, we will illustrate how certain unusual features of some well-known examples of the Schrödinger equation (and one less well-known) suggest underlying symmetries. We will also mention the Dirac equation and briefly highlight those features which also lead to a supersymmetric quantum mechanical interpretation. In Chapter 2, we will work out the examples the traditional way, by direct solution of the Schrödinger equation. In later chapters, these same examples will be revisited, and solved more elegantly, to illustrate the ideas of supersymmetric quantum mechanics.

But first and foremost, let us replace the cumbersome "supersymmetric quantum mechanics" with the acronym SUSYQM, pronounced "Suzy-Cue-Em."

The time dependent Schrödinger equation is

$$H(\mathbf{r}, t)\Psi(\mathbf{r}, t) \equiv -\frac{\hbar^2}{2m}\,\nabla^2\Psi(\mathbf{r}, t) + V(\mathbf{r}, t)\,\Psi(\mathbf{r}, t) = i\hbar\,\frac{\partial}{\partial t}\Psi(\mathbf{r}, t) \quad (1.1)$$

with $H(\mathbf{r}, t)$ the hamiltonian of the system.

All of the problems which we shall consider will involve time-independent one-dimensional potentials: either $V(x)$ in Cartesian coordinates, or $V(r)$, a radially symmetric potential in spherical coordinates. For any time-independent potential $V(\mathbf{r})$, the time dependence and space dependence can be separated:

$$\Psi(\mathbf{r}, t) = \psi(\mathbf{r})f(t) \ .$$

This yields, after a bit of work,

$$-\frac{\hbar^2}{2m}\frac{\nabla^2\psi(\mathbf{r})}{\psi(\mathbf{r})} + V(\mathbf{r}) = i\hbar\,\frac{df(t)/dt}{f(t)} = E \qquad (1.2)$$

where E is a constant, since $\mathbf{r}$ and t are independent variables.

Problem 1.1. *Obtain Eq. (1.2).*

Solving the time dependent part

$$\frac{i\hbar}{f(t)}\frac{df(t)}{dt} = E$$

we obtain[4] $f(t) = e^{-iEt/\hbar}$, while the space-dependent part gives the time-independent Schrödinger equation:

$$H(\mathbf{r})\psi(\mathbf{r}) \equiv -\frac{\hbar^2}{2m}\nabla^2\psi(\mathbf{r}) + V(\mathbf{r})\psi(\mathbf{r}) = E\psi(\mathbf{r}). \qquad (1.3)$$

From the wave-mechanical definition of linear momentum: $\mathbf{p} \equiv -i\hbar\nabla$, the first term on the left is the quantum extension of the kinetic energy $p^2/2m$, the second term is the extension of the potential energy; thus, E must be the total energy of the system. In one dimension,

$$-\frac{\hbar^2}{2m}\frac{d^2\psi(x)}{dx^2} + V(x)\psi(x) = E\psi(x). \qquad (1.4)$$

Let us begin by summarizing the results for the infinite square well in one dimension. The energy eigenvalues are found to be

$$E_n = n^2\hbar^2/2mL^2 \qquad n = 1, 2, 3, \ldots,$$

where m is the particle's mass and L is the width of the well.

Now let us look at another problem, far less well known: the potential $V(x) = \frac{\hbar^2}{2m}\left(2\,\text{cosec}^2\frac{\pi x}{L} - 1\right)$. Why we chose this particular form will become evident when we compare it to the infinite well. The energy eigenvalues turn out to be

$$E_n = [(n+2)^2 - 1]\hbar^2/2mL^2 \qquad n = 0, 1, 2, \ldots .$$

We notice something remarkable about the energy spectrum. It is virtually identical to that of the infinite well, except for the starting value of $n = 0$ rather than 1, the shift from n^2 to $(n+2)^2$, and the constant shift -1. These distinctions are independent of the physics: we could shift the infinite well's bottom from $V = 0$ to $V = -\frac{\hbar^2}{2mL^2}$; we could just as easily

---

[4]We take the multiplicative constant equal to 1, absorbing normalization into the spacial part of the wave function.

start the infinite well count at $n = 0$. Then the energy spectrum for the infinite well would be given by

$$E_n = [(n+1)^2 - 1]\hbar^2/2mL^2 \qquad n = 0, 1, 2, \ldots.$$

Thus, for some reason these two very different potentials yield virtually identical spectra[5]. Surely this cannot be coincidental. In fact, it is a signal that there is a hidden symmetry connecting the two problems.

Our third example is the harmonic oscillator: $V(x) = \frac{1}{2} m\omega^2 x^2$. The Schrödinger equation becomes

$$-\frac{\hbar^2}{2m}\frac{d^2\psi(x)}{dx^2} + \frac{1}{2}\omega^2 x^2\psi(x) = E\psi(x) . \qquad (1.5)$$

The energy turns out to be

$$E_n = \left(n + \frac{1}{2}\right)\hbar\omega .$$

The full solution of the Schrödinger equation, which we shall work out in Chapter 2, is quite cumbersome. This is surprising, given how symmetric the hamiltonian is: quadratic in both momentum $p = -i\hbar\frac{d}{dx}$ and position x. It would be nice if we then could factor it into two linear terms. Unfortunately, the fact that x and $\frac{d}{dx}$ do not commute seems to thwart this program. But as Fock found, this factorization approach is more powerful than the direct solution of the Schrödinger equation. Indeed, it will prove to be the first step toward SUSYQM. We shall work it out in the Chapter 4.

Let us summarize one more standard example, this time for a three-dimensional potential $V(r)$: the hydrogen atom. For this case, the standard treatment is in spherical coordinates. Since the Coulomb potential $V(r) = \frac{-e^2}{4\pi\epsilon_0 r}$ is independent of angle, the solution is of the form $\psi(r, \theta, \phi) = R(r)\Theta(\theta)\Phi(\phi)$, where only $R(r)$ depends on the potential. With the substitution $u(r) = R(r)/r$ the radial equation reduces to

$$-\frac{\hbar^2}{2m}\frac{d^2u(r)}{dr^2} + \left[\frac{-e^2}{4\pi\epsilon_0 r} + \frac{\hbar^2}{2m}\frac{l(l+1)}{r^2}\right]u(r) = Eu(r) , \qquad (1.6)$$

where l is the angular momentum quantum number. The energies are found to be $E_n = -\left[\frac{m}{2\hbar^2}\left(\frac{-e^2}{4\pi\epsilon_0}\right)^2\right]/n^2 = -13.6 \ eV/n^2$ in atomic units.

We notice something interesting here. The potential had a dependence on the angular momentum quantum number l, but the energy depends only on the principal quantum number n. That is, states of different potentials can have the same energy. This is a feature we will encounter often

---

[5]Except there is one extra state for the infinite well, for $n = 0$.

as we consider the supersymmetric approach. Indeed, we might say that SUSYQM reverses the usual quantum mechanical paradigm. Instead of asking, "For a given potential, what are the allowed energies it produces?" SUSYQM asks, "For a given energy, what are the potentials that could have produced it?"

We have seen that the infinite well and the $cosec^2$ potential yield similar energy spectra. They will turn out to be each others' supersymmetric partner potentials. This, as we shall see, will provide us with a way for generating the eigenfunctions of one potential from those of the other, *provided* that we know one of them already. The harmonic oscillator is a special case, partnering with itself, and generating a single set of eigenfunctions. Its importance here resides in its being the first and simplest application of the factorization method which leads to SUSYQM. The hydrogen atom has provided us with a three-dimensional example (among many) which will prove to be accessible to the SUSYQM algebraic approach.

Next we will introduce the concept of "shape invariance," which will allow us to obtain the complete energy spectrum of a single potential, without knowing *a priori* the spectrum of its partner. *We will do all of this without ever directly solving the Schrödinger differential equation.*

We mentioned earlier that SUSYQM is not restricted to the Schrödinger equation. We shall examine the Dirac equation in Chapter 15. Let us simply note at this point that, keeping the definition $\mathbf{p} \equiv -i\hbar\nabla$, Dirac's relativistic extension of the time-dependent wave equation for a particle in a scalar potential is

$$H\Psi \equiv \left[c\alpha \cdot \mathbf{p} + \beta mc^2 + V(\mathbf{r}, t) \right] \Psi = i\hbar\partial\Psi/\partial t.$$

α and β turn out to be not scalars, but 4×4 matrices. This requires that the equation have as its solution, not a single function Ψ, but rather a four-component column matrix constructed of such functions:

$$\begin{pmatrix} \Psi_1 \\ \Psi_2 \\ \Psi_3 \\ \Psi_4 \end{pmatrix}.$$

The equations coupling these Ψ_i's to each other can be rendered into the same form as the equations which generate the eigenfunctions of one Schrödinger potential from those of its partner. The Dirac equation, we might therefore say, is intrinsically supersymmetric.

Chapter 2

Traditional Quantum Mechanics — Clues to SUSYQM

In this chapter, we shall work out the details of the four examples which we mentioned in the previous chapter: the infinite well, the $cosec^2$ potential (also called Pöschl-Teller I, or trigonometric Pöschl-Teller potential), the harmonic oscillator, and the hydrogen atom. We shall do this by direct solution of the Schrödinger equation. We want to demonstrate the complexity (and tediousness) of this standard approach, so that it may be compared with the more elegant factorization method of supersymmetry in subsequent chapters.

Quick Facts Review

- The time-independent Schrödinger equation in one dimension is

$$-\frac{\hbar^2}{2m}\frac{d^2\psi(x)}{dx^2} + V(x)\psi(x) = E\psi(x) . \qquad (2.1)$$

- $|\psi(x)|^2 dx$ represents the probability of finding a particle between x and $x + dx$. If we consider only those cases where particles do not decay, the probability of finding them somewhere must be unity. I.e., the wave function is normalized so that

$$\int_{-\infty}^{\infty} |\psi(x)|^2 dx = 1 . \qquad (2.2)$$

- In addition, $\psi(x)$ must be continuous over its entire range. Were it not, then the probability $|\psi(x)|^2 dx$ of finding a particle of a given energy between x and $x + dx$ would not be single-valued, and thus ill-defined. We also usually insist that $\frac{d\psi(x)}{dx}$ be continuous, in order that the term $\frac{d^2\psi(x)}{dx^2}$ in the Schrödinger equation be well-defined. This constraint is suspended when $V(x)$ has a sharp discontinuity.

Now let us turn to our four examples.

Example 1: The Infinite Square Well, or "Particle in a Box"

Let us take the box to stretch from 0 to L. Then

$$V(x) = \begin{cases} \infty, & x < 0 \\ 0, & 0 \le x \le L \\ \infty, & x > L \end{cases}.$$

Outside the well, $V(x) = \infty$ requires that $\psi(x) = 0$. Inside the well, since the potential is 0,

$$-\frac{\hbar^2}{2m}\frac{d^2\psi(x)}{dx^2} = E\psi(x). \tag{2.3}$$

The solution is $\psi(x) = A \sin kx + B \cos kx$, where $k \equiv \sqrt{2mE}/\hbar$.

The wave function must be continuous in order to insure that the probability is meaningful across a barrier [1]; thus, ψ must vanish at the boundaries. For $x = 0$ we have $\psi(0) = A \sin 0 + B \cos 0 = 0$, which implies $B = 0$; viz., $\psi(x) = A \sin kx$. Then at the other boundary, $\psi(L) = A \sin kL = 0$. This means that $kL = n\pi$, where $n = 1, 2, 3...$; i.e., $k = n\pi/L$, thus quantizing the energy:

$$E = \hbar^2 k^2/2m = n^2\pi^2\hbar^2/2mL^2 .$$

The wave function is

$$\psi(x) = A \sin(n\pi x/L) .$$

Since $\psi(x) = 0$ outside the well, the normalization integral need only be taken from 0 to L:

$$\int_0^L |\psi(x)|^2 dx = \int_0^L A^2 \sin^2(n\pi x/L) dx = 1 .$$

Thus, $A = \sqrt{2/L}$.

Summarizing, and subscripting the wave function and energy:

$$\psi_n(x) = \sqrt{\frac{2}{L}} \sin\left(\frac{n\pi x}{L}\right) \tag{2.4}$$

$$E_n = \frac{n^2\pi^2\hbar^2}{2mL^2} . \tag{2.5}$$

Now let us make some useful simplifications. The first simplification is to choose physical units such that $2m = 1$ and $\hbar = 1$. (This we will continue

---

[1]Its derivative however is not continuous, since $V(x)$ is sharply discontinuous at $x = 0$ and $x = L$.

to do throughout the book, unless we specify otherwise.) The second is to choose a length unit for the well, $L = \pi$. Then our solutions for the wave function and energy simplify greatly:

$$\psi_n(x) = \sqrt{\frac{2}{\pi}} \, \sin nx. \tag{2.6}$$

$$E_n = n^2 \qquad n = 1, 2, 3, \dots \tag{2.7}$$

The potential and first three wave functions and the corresponding eigenenergies are illustrated in Figure 2.1.

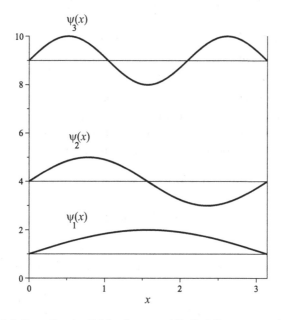

Fig. 2.1   The infinite well potential for $L = \pi$ and its first three energy levels and wave functions.

Example 2: The Trigonometric Pöschl-Teller Potential

The potential is given by

$$V(x) = 2\mathrm{cosec}^2 x - 1 \, , \;\; 0 < x < \pi \, . \tag{2.8}$$

To solve the corresponding Schrödinger equation [2]

$$-\frac{d^2\psi(x)}{dx^2} + \left(2\,\mathrm{cosec}^2 x - 1\right)\psi(x) = E\psi(x), \qquad (2.9)$$

we change variables to $y = \sin^2\left(\frac{x}{2}\right)$ and $\psi(x(y)) = \frac{(-1+y)}{\sqrt{y}}\,\phi(y)$. While the domain of x is given by $(0, \pi)$, it is now mapped to $(0, 1)$, i.e., $(0 < y < 1)$. The new wave function $\phi(y)$ is related to our original wave function by

$$\phi(y) = -\frac{\sin\left(\frac{x}{2}\right)}{\cos^2\left(\frac{x}{2}\right)}\;(x)\;.$$

These substitutions lead to:

$$y(1-y)\frac{d^2\phi(y)}{dy^2} + \left(-\frac{1}{2} - 2y\right)\frac{d\phi(y)}{dy} - \left(-\frac{3}{4} - E\right)\phi(y)\;. \qquad (2.10)$$

This equation is the hypergeometric equation [3]:

$$y(1-y)\frac{d^2\phi(y)}{dy^2} + (c - (a+b+1)y)\frac{d\phi(y)}{dy} - ab\,\phi(y) = 0\;. \qquad (2.11)$$

Its solutions are the hypergeometric functions $_2F_1(a, b, c, y)$. Comparing Eqs. (2.10) and (2.11), we identify $a = \frac{1}{2} - \sqrt{1+E}$, $b = \frac{1}{2} + \sqrt{1+E}$, and $c = -\frac{1}{2}$. The general solution of this equation, consisting of a linear combination of two independent functions, is given by

$$\phi(y) = \alpha\; _2F_1\left(\frac{1}{2} - k, \frac{1}{2} + k, -\frac{1}{2}, y\right) \qquad (2.12)$$

$$+ \beta\, y^{\frac{3}{2}}\; _2F_1\left(2 - k, 2 + k, \frac{5}{2}, y\right)$$

where $k = \sqrt{1+E}$. Thus, our original wave function $\psi(x(y)) = \frac{(-1+y)}{\sqrt{y}}\,\phi(y)$, which we shall simply call $\psi(y)$, is now given by

$$\psi(y) = \alpha\,\frac{(y-1)}{\sqrt{y}}\; _2F_1\left(\frac{1}{2} - k, \frac{1}{2} + k, -\frac{1}{2}, y\right) \qquad (2.13)$$

$$+ \beta\, y(y-1)\; _2F_1\left(2 - k, 2 + k, \frac{5}{2}, y\right)$$

---

[2] At this point, we would like to caution especially our undergraduate readers, that the traditional method of solving the Schrödinger equation, which we shall now carry out, leads to a general solution involving hypergeometric functions, which may not be familiar to you. If you do not find it very instructive at this stage, feel free to skip the details of this example at this time. Do, however, look at the final answers. In Chapter 5 we will again solve this problem using the much more accessible methods of SUSYQM. You may revisit this example at that time for a comparative study of the two methods.

[3] For a nice discussion of this, see Arfken and Weber, *Mathematical Methods for Physicists*, 5th edition, San Diego: Harcourt-Academic Press (2001).

where α and β are constants to be determined. The first term in this solution blows up at $y = 0$ (i.e., at $x = 0$), which will yield an unphysical infinite probability. So the only way this solution can be well behaved is if $\alpha = 0$. With this choice, $\psi(y)$ reduces to

$$\psi(y) = y(y-1) \, _2F_1\left(2-k, \, 2+k, \, \frac{5}{2}, \, y\right).$$

We wish to apply the boundary condition $\psi(x(y)) = 0$ at $y = 1$, where $V \to \infty$. But we cannot immediately do so, as the hypergeometric function becomes infinite at $y = 1$. We need to investigate the behavior of the entire function $\psi(y) = \frac{(-1+y)}{\sqrt{y}}\phi(y)$. We use an identity that extracts the essence of this singularity,

$$_2F_1\left(a, b, c, y\right) = \frac{\Gamma\left(c\right)\Gamma\left(c-a-b\right)}{\Gamma\left(c-a\right)\Gamma\left(c-b\right)} \, _2F_1\left(a, b, a+b-c+1, 1-y\right) \quad (2.14)$$

$$+ (1-y)^{c-a-b} \frac{\Gamma\left(c\right)\Gamma\left(a+b-c\right)}{\Gamma\left(a\right)\Gamma\left(b\right)} \, _2F_1\left(c-b, c-a, c-a-b+1, 1-y\right).$$

Γ denotes the Gamma function[4]. These hypergeometric functions are no longer singular when $y \to 0$ as the argument $(1-y)$ goes to zero. The singularity is contained in the term $(1-y)^{c-a-b}$. Thus, for $\psi(y)$, we obtain

$$\psi(y) = y(y-1) \frac{\Gamma\left(\frac{5}{2}\right)\Gamma\left(-\frac{3}{2}\right) \, _2F_1\left(2-k, 2+k, \frac{5}{2}, 1-y\right)}{\Gamma\left(\frac{1}{2}-k\right)\Gamma\left(\frac{1}{2}+k\right)} \quad (2.15)$$

$$-y\left(1-y\right)^{-\frac{1}{2}} \frac{\Gamma\left(\frac{5}{2}\right)\Gamma\left(\frac{3}{2}\right) \, _2F_1\left(\frac{1}{2}-k, \frac{1}{2}+k, -\frac{1}{2}, 1-y\right)}{\Gamma\left(2+k\right)\Gamma\left(2-k\right)}.$$

The second term is infinite at $y = 1$ due to $1-y$ in the denominator. The hypergeometric function is 1 when its variable argument, in this case $(1-y)$, is zero. For this wave function to be well behaved at $x = \pi$, therefore, the second term must vanish. This is possible only if

$$\Gamma\left(2+k\right)\Gamma\left(2-k\right) \to \infty,$$

which occurs when the arguments of $\Gamma\left(2+k\right)$, or $\Gamma\left(2-k\right)$ equal a negative integer or zero. Thus, we set $2\pm k \equiv 2\pm\sqrt{1+E} = -n$, where $n = 0, 1, 2, \ldots$, thus quantizing the energy [5].

---

[4]$\Gamma(x)$ behaves like $\frac{(-1)^n}{n!}\frac{1}{x+n}$ near $x = -n$, $n = 0, 1, 2, \ldots$. For positive integer n, $\Gamma(n+1) = n!$.

[5]Careful analysis, which we gloss over, shows that the same condition that makes the second term in Eq. (2.15) vanish also truncates the first term to a polynomial of order n. Otherwise the wave functions would diverge.

Solving for E, we get:

$$E_n = (n+2)^2 - 1 \qquad n = 0, 1, 2, \ldots. \tag{2.16}$$

If we investigate the hypergeometric function for, say, the lowest two states: $n = 0$ and 1, we find that it simplifies, yielding $\psi_0(x) \sim \sin^2 x$ and $\psi_1(x) \sim \sin x \, \sin 2x$.

Figure 2.2 shows the potential and the first two energy levels and wave functions. Note the complicated effort expended to obtain what appear to be rather simple wave functions, and a very simple expression for the energies.

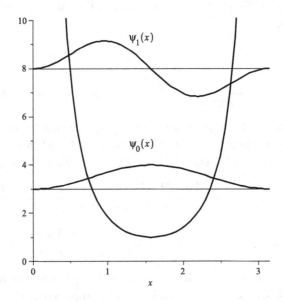

Fig. 2.2   The $2 \cosec^2 x - 1$ potential and its first two energy levels and wave functions.

If we now go back to the infinite well spectrum, Eq. (2.7), have it start at $n = 0$ instead of $n = 1$, and shift its bottom to $V(x) = -1$ instead of 0, neither of which affects the physics, then its energy spectrum is given by

$$E_n = (n+1)^2 - 1 \qquad n = 0, 1, 2, \ldots. \tag{2.17}$$

This means that the infinite well and the $cosec^2$ potentials yield *identical* spectra, except for the infinite well's extra state: its ground state $n = 0$, with energy E_0 now equal to 0. Why this "coincidence" occurs will be revealed by SUSYQM.

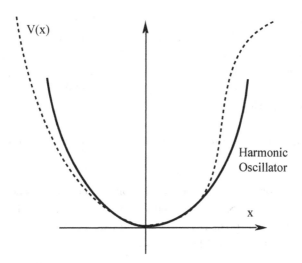

Fig. 2.3 For small x-regions near the minimum the harmonic oscillator potential approximates any arbitrary attractive potential $V(x)$.

Example 3: The Harmonic Oscillator

The harmonic oscillator ($\frac{1}{2} m\omega^2 x^2$ in ordinary units), is arguably the most important potential in quantum mechanics. It approximates any attractive potential in a narrow region near the minimum, as illustrated in Figure 2.3. When we consider realistic systems, we observe that any smooth attractive potential $V(x)$ has a minimum corresponding to the equilibrium position. In this small region around the minimum, the potential can be approximated by its Taylor series expansion

$$V(x) = V(0) + \left.\frac{dV}{dx}\right|_{x=0} x + \frac{1}{2} \left.\frac{d^2V}{dx^2}\right|_{x=0} x^2 + \dots \qquad (2.18)$$

where x is the distance from the minimum. That minimum, the equilibrium position, has $dV/dx|_{x=0} = 0$. We can always set the constant term $V(0) = 0$, since this does not change the physics. Thus, the potential energy of any such system can be approximated as

$$V(x) = \frac{1}{2} k x^2 \qquad (2.19)$$

where we define $\left.\frac{d^2 V}{d x^2}\right|_{x=0} \equiv k = m\omega^2$. This then is the harmonic oscillator potential, which, in our convention ($\hbar = 1,\ 2m = 1$), becomes

$$V(x) = \frac{1}{4} \omega^2 x^2 \ . \qquad (2.20)$$

Finding the exact solutions for the harmonic oscillator potential will give us insight into the solutions of more complicated potentials. But, as we shall see in the next chapter, the harmonic oscillator is a very important system in its own right. Its form hints at a deeper symmetry of the hamiltonian, one that will be crucial in developing the SUSYQM formalism. Here we simply work out the problem by brute force solution of the Schrödinger equation.

Using our simplified units, the Schrödinger equation becomes

$$-\frac{d^2\psi(x)}{dx^2} + \frac{1}{4}\,\omega^2\,x^2\psi(x) = E\psi(x) \ . \tag{2.21}$$

Let us first analyze its asymptotic solutions at large x. When $|x| \to \infty$, the potential $V(x) \to \infty$ as well, and we expect two things: that the wave functions $\psi(x)$ will vanish (as we saw them do for the infinite well), and that the energy term E will be negligible with respect to the x^2 term. The latter condition means that at large $|x|$, Eq. (2.21) becomes

$$\frac{d^2\psi(x)}{dx^2} \approx \frac{1}{4}\,\omega^2\,x^2\psi(x) \ , \tag{2.22}$$

whose approximate solution is the Gaussian function $e^{-\omega x^2/4}$.

Problem 2.1. *Show that $\psi(x) = e^{-\omega x^2/4}$ is an approximate solution of Eq. (2.22).*

We observe that indeed, the Gaussian wave function vanishes at infinity, in agreement with our physical intuition. We can now try a "peeling-off" process, as follows: we choose an ansatz [6] for the general solution of Eq. (2.21)

$$(x) = C\,e^{-\omega x^2/4}\,v(x) \ , \tag{2.23}$$

where C is a normalization constant, to be determined later. Thus, we have "peeled off" the Gaussian, and need only find the remaining function $v(x)$. We must keep in mind, however, that $v(x)$ cannot rise faster than $e^{-\omega x^2/4}$ falls, as $x \to \pm\infty$, otherwise $\psi(x)$ will diverge.

Substituting Eq. (2.23) into (2.21) and making a simplifying change of variables

$$\xi \equiv x\,\sqrt{\omega/2} \ , \quad K \equiv 2\,E/\omega \ , \tag{2.24}$$

we obtain

$$\frac{d^2v(\xi)}{d\xi^2} - 2\,\xi\,\frac{dv(\xi)}{d\xi} + (K-1)\,v(\xi) = 0 \ , \tag{2.25}$$

Problem 2.2. *Do the explicit calculations to derive Eq. (2.25).*

---

[6]An ansatz is a guess as to the form of the solution.

We can solve this differential equation using the series method. We express the function $v(x)$ by a power series

$$v(\xi) = \sum_{j=0}^{\infty} v_j \, \xi^j \tag{2.26}$$

where the the coefficients v_j are to be chosen such that the equation (2.25) is identically satisfied for each power of ξ. We now need to substitute Eq. (2.26) into Eq. (2.25). We obtain $dv(\xi)/d\xi = \sum_{j=0}^{\infty} j v_j \xi^{j-1}$, whence $2\xi v(\xi) = 2\sum_{j=0}^{\infty} j v_j \xi^j$. (The first term is 0, but that has no effect on the calculation.) For the second derivative, we obtain $d^2 v(\xi)/d\xi^2 = \sum_{j=0}^{\infty} j(j-1)v_j \xi^{j-2}$. We observe that the first two terms of this series are 0, so we may rewrite it starting with the third term: $\sum_{j=2}^{\infty} j(j-1)v_j \xi^{j-2}$ or, with a change of index, $\sum_{j=0}^{\infty}(j+2)(j+1)v_{j+2}\xi^j$. Now all three terms in Eq. (2.25) have the same powers of ξ:

$$\sum_{j=0}^{\infty} \left[(j+2)(j+1)v_{j+2} - 2jv_j + (K-1)v_j \right] \xi^j = 0. \tag{2.27}$$

This can be satisfied if and only if all the coefficients of ξ^j vanish. This leads to the following recursion relation:

$$v_{j+2} = \frac{2j+1-K}{(j+1)(j+2)} \, v_j. \tag{2.28}$$

Note that this relation separates the solutions into those of odd and even parity. Given the initial coefficient v_0 we can find all even v_j; given v_1 we can calculate all odd v_j. If $v_0 = 0$ then from Eq. (2.28) we obtain that $v_2 = v_4 = \ldots = 0$. Then in the series expansion for v, Eq. (2.26), we are left with only the odd powers of ξ : $v(\xi) = v_1\xi + v_3\xi^3 + \ldots$. Similarly, if $v_1 = 0$, then $v_3 = v_5 = \ldots = 0$, and the series expansion of v will contain only even terms: $v(\xi) = v_0 + v_2\xi^2 + v_4\xi^4 + \ldots$.

It would seem that a general solution could be a mix of odd and even terms, provided we knew *a priori* the coefficients v_0 and v_1. But before we worry about this, we need to see if each of these series converges, and if so, to what. In particular, we must assure that when multiplied by the peeled-off Gaussian $e^{-\omega x^2/4}$, $v(x)$ gives a normalizable $\psi(x) = C e^{-\omega x^2/4} v(x)$.

For $j \gg 1$ and K, the recursion relation Eq. (2.28) may be approximated by

$$v_{j+2} \approx \frac{2}{j} v_j. \tag{2.29}$$

First we need to establish that the approximate series is an upper bound of the exact series, Eq. (2.28), for if it isn't, then we would not know if the exact series gives a normalizable ψ, even if the approximate series does. We can in fact show that the approximate coefficients $\frac{2}{j}$ exceed the exact coefficients $\frac{2j+1-K}{(j+1)(j+2)}$ for every j. Since the potential well's bottom is at 0, all energies must be positive; thus, $K \equiv 2E/\omega > 0$. Consequently

$$\frac{2j+1-K}{(j+1)(j+2)} < \frac{2j+1}{(j+1)(j+2)} < \frac{2j+2}{(j+1)(j+2)} = \frac{2}{j+2} < \frac{2}{j}.$$

Thus the approximate series overestimates the true series. So if the approximate series converges so that ψ is normalizable, we are assured that the correct series will converge. The coefficients which obey the approximate recursion relation are $v_j \approx C/(j/2)!$.

Problem 2.3. *Confirm that this relation yields Eq. (2.29).*

Then the approximate series is $v(\xi) \approx C \sum_{j=0}^{\infty} \frac{1}{(j/2)!} \xi^j$. Defining $j' \equiv 2j$, $v(\xi) \approx C \sum_{j'=0}^{\infty} \frac{1}{j'!} \xi^{2j'} \approx Ce^{\xi^2}$.

So unfortunately the approximate series converges to $e^{\xi^2} \equiv e^{\omega x^2/2}$, which overwhelms the decaying Gaussian $e^{-\omega x^2/4}$, producing a divergent wave function at $\pm\infty$.

We need for the exact series to converge to $e^{-\omega x^2/n}$, $n > 4$, which we showed is less than the approximate series. When multiplied by the decaying Gaussian, this would produce a normalizable wave function. But our approximation must be good at every ξ. In particular, it must hold for very large ξ. Now note from Eqs. (2.27) and (2.28) that the ratio of successive exact terms $v_{j+2}\xi^{j+2}/v_j \xi^j = \frac{2j+1-K}{(j+1)(j+2)}\xi^2$. Thus, as $\xi \to \infty$, the larger j terms increasingly overwhelm the smaller ones, making the large-j approximate series more and more accurate. This means that the approximate and exact series converge to each other as $\xi \to \infty$. Therefore, the exact series behaves as badly as its approximation.

The only way out of this is to make sure that the exact series will always be overwhelmed by the decaying Gaussian. We accomplish this by truncating the series at some finite index n.

This can be done if in Eq. (2.28) we require that at some $j_{max} \equiv n$, $v_{j_{max}+2} = 0$. This will be the case if

$$K = 2n + 1 . \tag{2.30}$$

We note now that the value of $n \equiv j_{max}$ will be odd or even, depending on whether the series is odd or even. Running the process backwards, an odd n means a starting value $j = 1$, while an even n means a starting value $j = 0$. Thus, for example, an odd n leaves the even series untruncated,

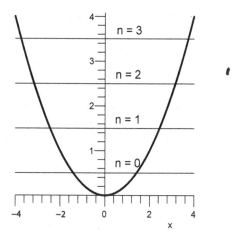

Fig. 2.4   The first four energy levels of the harmonic oscillator for $\omega = 1$.

converging to the prohibited $e^{\omega x^2/2}$, and conversely for even n. So the solutions for a given energy must be of either odd or even parity, with the leading term of the other parity, v_0 or v_1, set to 0.

From Eq. (2.24) we see that the constraint $K = 2n + 1$ quantizes the energy:

$$E_n = \left(n + \frac{1}{2} \right) \omega \ , \quad n = 0, 1, 2, \ldots . \tag{2.31}$$

The energy levels are equally spaced by ω (in $\hbar = 1$ units) as illustrated in Figure 2.4.

We now know that the solution to the Eq. (2.21) is given by the product of the Gaussian function $e^{-\omega x^2/4}$ and the solution of Eq. (2.25) when we substitute for K the value $2n + 1$. Equation (2.25) thus becomes

$$\frac{d^2 v(\xi)}{d\xi^2} - 2\xi \frac{dv(\xi)}{d\xi} + 2n\, v(\xi) = 0 \ , \quad n = 0, 1, 2, \ldots . \tag{2.32}$$

This is the Hermite equation whose solutions are the Hermite polynomials. The first four polynomials are

$$H_0(\xi) = 1, \ H_1(\xi) = 2\xi, \ H_2(\xi) = 4\xi^2 - 2, \ H_3(\xi) = 8\xi^3 - 12\xi . \tag{2.33}$$

Hermite polynomials can be generated directly using the Rodrigues differential formula

$$H_n(\xi) = (-1)^n\, e^{\xi^2} \frac{d^n}{d\xi^n} \left(e^{-\xi^2} \right) . \tag{2.34}$$

Using this, we can show by direct manipulation that they have the following properties

$$H'_n = 2nH_{n-1} \, , \ \ H_{n+1} - 2\xi H_n + 2nH_{n-1} = 0 \, , \ \ H_{n+1} = 2\xi H_n - H'_n \, , \quad (2.35)$$

where the prime represents the derivative with respect to ξ.

Problem 2.4. *Use the Rodrigues formula to confirm the values of the first four Hermite polynomials, H_0 to H_3, in Eq. (2.33), and to find the fifth, H_4. Test that the recursion relations Eq. (2.35) are obeyed by the polynomials in Eq. (2.33).*

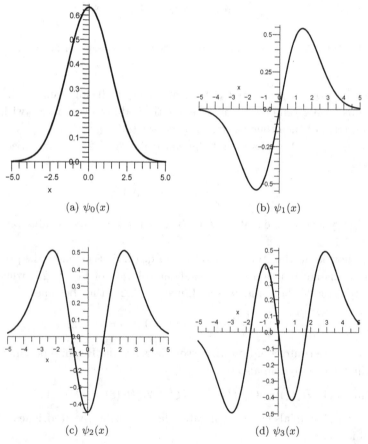

(a) $\psi_0(x)$ (b) $\psi_1(x)$

(c) $\psi_2(x)$ (d) $\psi_3(x)$

Fig. 2.5   Normalized harmonic-oscillator wave functions for $\omega = 1$ and quantum numbers $n = 0, 1, 2, 3$.

The constant C from Eq. (2.23) can be determined from the normalization condition $\int_{-\infty}^{\infty} |\psi_n|^2 \, dx = 1$. We obtain $C \equiv C_n = \left(\frac{\omega}{2}\right)^{\frac{1}{4}} [\sqrt{\pi}\, n!\, 2^n]^{-\frac{1}{2}}$.

Putting these results together, we conclude that the wave function $\psi_n(x)$ and the corresponding energy E_n for the harmonic oscillator are given by

$$\psi_n(x) = \left(\frac{\omega}{2}\right)^{\frac{1}{4}} [\sqrt{\pi}\, n!\, 2^n]^{-\frac{1}{2}}\, e^{-\omega x^2/4} H_n(x\sqrt{\omega/2}) \; ,$$

$$E_n = \left(n + \frac{1}{2}\right) \omega \; , \quad n = 0, 1, 2, \ldots . \tag{2.36}$$

The eigenfunctions $\psi_n(x)$ for $n = 0, 1, 2, 3$ are shown in Figure 2.5.

Before we conclude this section we should make one very important observation. In Eq. (2.35) we saw that the Hermite polynomials obey certain recursion relations. These relations, in terms of the wave functions ψ_n, become:

$$\sqrt{n+1}\, \psi_{n+1} - \sqrt{2\omega}\, x\, \psi_n + \sqrt{n}\, \psi_{n-1} = 0 \; , \tag{2.37}$$

$$\psi'_n + \frac{1}{\sqrt{2}}\, \omega\, x \psi_n = \sqrt{n\omega}\, \psi_{n-1} \; , \tag{2.38}$$

$$-\psi'_n + \frac{1}{\sqrt{2}}\, \omega\, x\, \psi_n = \sqrt{(n+1)\,\omega}\, \psi_{n+1} \; , \tag{2.39}$$

where prime represents the derivative with respect to x. That is, some rather simple operations relate ψ_n to its neighbors ψ_{n-1} and ψ_{n+1}.

Problem 2.5. *Use the recursion relations for the Hermite polynomials, Eq. (2.35), to prove Eqs. (2.37), (2.38) and (2.39).*

The simplicity of the solution for the infinite well seemed commensurate with the amount of work expended to obtain it. This is not true for the the harmonic oscillator (as was the case for the *cosec*^{2} problem.) The work seems quite excessive compared to the simplicity of the solutions. There is an easier way: the so-called factorization method to which we alluded in the introduction. It provided the first example of what subsequently led to supersymmetric quantum mechanics. In the next chapter, we shall work out the harmonic oscillator problem using the factorization method.

Problem 2.6. *We will soon see that we always want to design potentials in SUSYQM so that the ground state energy is 0. (We did this for the infinite well in Eq.(2.17)). Modify the Schrödinger equation for the harmonic oscillator, Eq. (2.21), so that $E_0 = 0$. Argue that this does not change the wave functions.*

Problem 2.7. *Use your results from the infinite well and the harmonic oscillator to obtain the wave functions and energies for the "half-harmonic oscillator":*

$$V(x) = \begin{cases} \infty & , \ x < 0 \\ \frac{1}{4}\omega^2 x^2 & , \ x \geq 0 \end{cases}.$$

(Hint: the well shape is that of the harmonic oscillator; the infinite wall constitutes a boundary condition on the wave functions. You can get the answer by just considering the symmetries of the problem.)

Example 4: The Hydrogen Atom

The hydrogen atom problem is also known as the Coulomb problem. The standard treatment is in spherical coordinates. The time-independent Schrödinger equation in three dimensions, given in the introduction, is

$$H(\mathbf{r})\psi(\mathbf{r}) \equiv -\nabla^2\psi(\mathbf{r}) + V(\mathbf{r})\psi(\mathbf{r}) = E\psi(\mathbf{r}) \qquad (2.40)$$

or,

$$H\psi \equiv \frac{1}{r^2}\frac{\partial}{\partial r}\left(r^2\frac{\partial}{\partial r}\right) + \frac{1}{r^2\sin\theta}\frac{\partial}{\partial\theta}\left(\sin\theta\frac{\partial\psi}{\partial\theta}\right) + \frac{1}{r^2\sin^2\theta}\left(\frac{\partial^2}{\partial\phi^2}\right) = E\psi. \qquad (2.41)$$

We choose the ansatz that the wave function can be separated into a product of functions of the respective variables:

$$\psi(r,\theta,\phi) = R(r)\Theta(\theta)\Phi(\phi).$$

Since the Coulomb potential [7] is $V(r) = -e^2/r$, independent of angle, we can write the angular dependence as a single function $Y(\theta,\phi)$, and the wave function as $\psi(r,\theta,\phi) = R(r)Y(\theta,\phi)$, where only $R(r)$ depends on the potential. The Schrödinger equation becomes

$$Y(\theta,\phi)\frac{1}{r^2}\frac{d}{dr}\left(r^2\frac{dR(r)}{dr}\right) + R(r)\frac{1}{r^2\sin\theta}\frac{\partial}{\partial\theta}\left(\sin\theta\frac{\partial Y(\theta,\phi)}{\partial\theta}\right) \qquad (2.42)$$

$$+ R(r)\frac{1}{r^2\sin^2\theta}\left(\frac{\partial^2 Y(\theta,\phi)}{\partial\phi^2}\right) = ER(r)Y(\theta,\phi).$$

Multiplying by $-r^2/R(r)Y(\theta,\phi)$ it separates the radial and angular parts:

$$\left[\frac{1}{R(r)}\frac{d}{dr}\left(r^2\frac{dR(r)}{dr}\right) + r^2\left(\frac{e^2}{r} + E\right)\right] \qquad (2.43)$$

$$+ \frac{1}{Y(\theta,\phi)}\left[\frac{1}{\sin\theta}\frac{\partial}{\partial\theta}\left(\sin\theta\frac{\partial Y(\theta,\phi)}{\partial\theta}\right) + \frac{1}{\sin^2\theta}\left(\frac{\partial^2 Y(\theta,\phi)}{\partial\phi^2}\right)\right] = 0.$$

[7] In our unit system, $4\pi\epsilon_0$ is also equal to 1.

Each of the terms must be constant. Presciently calling that constant $l(l+1)$, we obtain

$$\frac{1}{R(r)}\frac{d}{dr}\left(r^2\frac{dR(r)}{dr}\right) + r^2\left(\frac{e^2}{r} + E\right) = l(l+1) ; \tag{2.44}$$

$$\frac{1}{Y(\theta,\phi)}\left[\frac{1}{\sin\theta}\frac{\partial}{\partial\theta}\left(\sin\theta\frac{\partial Y(\theta,\phi)}{\partial\theta}\right) + \frac{1}{\sin^2\theta}\left(\frac{\partial^2 Y(\theta,\phi)}{\partial\phi^2}\right)\right] = -l(l+1). \tag{2.45}$$

The angular solutions, which are the same for all radially dependent V are the spherical harmonics $Y_l^m(\theta,\phi)$. We will not concern ourselves with them. Note that the radial equation alone determines the energy.

With the substitution $u(r) = R(r)r$ the radial equation reduces to

$$-\frac{d^2u}{dr^2} + \left[\frac{-e^2}{r} + \frac{l(l+1)}{r^2}\right]u = Eu . \tag{2.46}$$

Problem 2.8. *Check that this substitution yields Eq. (2.46) from Eq. (2.44).*

Note that the effective potential is composed of two terms: the Coulomb attraction $-e^2/r$, plus a centrifugal repulsion $l(l+1)/r^2$ which depends on the angular momentum quantum number l, signaling that this one-dimensional potential is actually the radial part of a three-dimensional potential.

The range of r is $(0,\infty)$. For bound states we take the energy to be 0 at ∞ and negative otherwise. We define $\kappa \equiv \sqrt{-E}$, and divide Eq. (2.46) by $-\kappa^2$. Separating the derivative term on the left side gives

$$\frac{1}{\kappa^2}\frac{d^2u}{dr^2} = \left[1 - \frac{e^2}{\kappa^2 r} + \frac{l(l+1)}{\kappa^2 r^2}\right]u. \tag{2.47}$$

We now make two additional simplifying substitutions: $\rho \equiv \kappa r$ and $\rho_0 \equiv e^2/\kappa$.

This yields

$$\frac{d^2u}{d\rho^2} = \left[1 - \frac{\rho_0}{\rho} + \frac{l(l+1)}{\rho^2}\right]u. \tag{2.48}$$

In a peel-off procedure analogous to that for the harmonic oscillator, we examine the approximate behavior of the equation at the asymptotes $\rho \to (0,\infty)$. As $\rho \to \infty$, the 1 dominates, giving

$$\frac{d^2u}{d\rho^2} \approx u. \tag{2.49}$$

The solution is $u \sim Ae^{-\rho} + Be^{\rho}$. We set $B = 0$; otherwise, u would blow up[8] as $\rho \to \infty$. So $u \sim Ae^{-\rho}$ at the large distance limit.

Now we consider the approximate behavior of u near the origin ($\rho \to 0$). In this case the dominant term is the centrifugal one: $\frac{l(l+1)}{\rho^2}$, for $l > 0$. (We shall consider the special case of $l = 0$ shortly.) In that case

$$\frac{d^2 u}{d\rho^2} \approx \frac{l(l+1)}{\rho^2} u. \tag{2.50}$$

The solution is $u \sim C\rho^{l+1} + D\rho^{-l}$.

Problem 2.9. *Derive the above solution of Eq. (2.50).*

We must set $D = 0$ in order that u not blow up at 0. So $u \sim C\rho^{l+1}$.

We are now ready for the peel-off ansatz: $u \sim \rho^{l+1} e^{-\rho} v(\rho)$. We substitute this into Eq. (2.48) and obtain, after some work, the following equation for v:

$$\rho \frac{d^2 v}{d\rho^2} + 2(l+1-\rho)\frac{dv}{d\rho} + [\rho_0 - 2(l+1)]v = 0. \tag{2.51}$$

Problem 2.10. *Work out the details, obtaining this equation by direct substitution of the ansatz into Eq. (2.48).*

We try a series expansion for v:

$$v = \sum_{j=0}^{\infty} v_j \rho^j .$$

Then $dv/d\rho = \sum_{j=0}^{\infty} j v_j \rho^{j-1}$, which, since the leading term is 0, is the same as $\sum_{j=0}^{\infty}(j+1)v_{j+1}\rho^j$.

We obtain the second derivative by differentiation of this last term:

$$\frac{d^2 v}{d\rho^2} = \sum_{j=0}^{\infty} j(j+1)v_{j+1}\rho^{j-1}.$$

Then Eq. (2.51) becomes

$$\rho \frac{d^2 v}{d\rho^2} + 2(l+1-\rho)\frac{dv}{d\rho} + [\rho_0 - 2(l+1)]v$$

$$= \sum_{j=0}^{\infty} \{j(j+1)v_{j+1} + (2l+1)(j+1)v_{j+1} - 2jv_j$$

$$+ [\rho_0 - 2(l+1)]\, v_j\}\, \rho^j = 0 . \tag{2.52}$$

---

[8]Note that $\int u^2 dr = \int R^2 r^2 dr$ and $\int |\psi|^2 d^3 r = \int Y_l^m(\theta, \phi)^2 \sin^2 \theta d\theta d\phi \int R^2 r^2 dr$. So if u blows up, then $\int |\psi|^2 d^3 r$ blows up, which would yield an infinite probability.

This can be satisfied if and only if all the coefficients of ρ^j vanish. This leads to the following recursion relation between the coefficients v_j:

$$v_{j+1} = \frac{2(j+l+1) - \rho_0}{(j+1)(j+2l+2)} v_j. \tag{2.53}$$

Now we need to examine the behavior of this recursion relation. We need to assure ourselves that $v(\rho)$ does not overwhelm ρ^{l+1} near 0, nor $e^{-\rho}$ near ∞, and thus produce a divergent probability.

The $\rho = 0$ limit is $u \sim \rho^{l+1} \sum_{j=0}^{\infty} v_j \rho^j$. Since both l and j are both positive, there is no possibility of a singularity for small ρ.

The $\rho = \infty$ limit *is* a problem. For large ρ, large values of j might produce a strongly divergent v, one that might dominate the decaying exponential $Ae^{-\rho}$, analogous to what we saw for the harmonic oscillator. We therefore examine the behavior of the recursion relation for large values of j. For $j \gg 1$ and l, Eq. (2.53) becomes

$$v_{j+1} \approx \frac{2}{j} v_j . \tag{2.54}$$

Problem 2.11. *Confirm that this relation yields Eq. (2.54).*

Problem 2.12. *In analogy with what we did for the harmonic oscillator, show that the large-j approximate series coefficients, Eq. (2.54), overestimate the exact series coefficients, Eq. (2.53).*

Substituting Eq. (2.54) into Eq. (2.52) we obtain

$$v \approx v_0 + \rho \sum_{j=1}^{\infty} \frac{2^{j-1}}{(j-1)!} v_0 \rho^{j-1} = v_0 \left[1 + \rho \sum_{k=0}^{\infty} \frac{(2\rho)^k}{(k)!} \right] \tag{2.55}$$

where $k \equiv j - 1$.

In this form, what happens at large ρ becomes evident:

$$v \approx v_0 \rho e^{2\rho}.$$

Then the function $u \sim \rho^{l+1} e^{-\rho} v(\rho) \rightarrow v_0 \rho^{l+2} e^{\rho}$, which blows up as $\rho \rightarrow \infty$.

The only way to eliminate this problem is to truncate the exact series at some j_{max}, such that $v_{j_{max}+1} = 0$. From Eq. (2.53) this can be accomplished by setting $2(j_{max} + l + 1) - \rho_0 = 0$. We define the "principal quantum number" $n \equiv j_{max} + l + 1$. (Note that this requires $l \leq n - 1$.)

Then the truncation requirement is $\rho_0 = 2n$. This gives us the quantized energy levels: from $\rho_0 \equiv e^2/\kappa$ and $\kappa \equiv \sqrt{-E}$, the energies are

$$E_n = -\kappa_n^2 = -\frac{e^4}{4n^2} . \tag{2.56}$$

(This equals $-13.6eV/n^2$ in atomic units.)

Problem 2.13. *Work out the details for Eq. (2.56).*

What about $l = 0$? In this case, Eq. (2.48) becomes

$$\frac{d^2u}{d\rho^2} = \left[1 - \frac{\rho_0}{\rho}\right] u. \tag{2.57}$$

In the $\rho \to 0$ limit, we may neglect the 1 on the right-hand side

$$\frac{d^2u}{d\rho^2} \approx -\frac{\rho_0}{\rho} u . \tag{2.58}$$

We are tempted to assume (as some texts do) that $u \sim \rho^{l+1}$ is fine for $l = 0$; viz., $u = C\rho$. If we succumb to this temptation, we find that the left hand side of Eq. (2.58) goes to 0 while the right hand side approaches a constant:$-C\rho_0$. Let us instead be more conservative, keeping the next-higher-order term in the series expansion: $u \sim C_1\rho + C_2\rho^2$. Then the left hand side of Eq. (2.58) becomes $2C_2$. The right hand side is $(-\frac{\rho_0}{\rho})(C_1\rho + C_2\rho^2) = -C_1\rho_0 + C_2\rho\rho_0$.

Equating right and left hand sides in the $\rho \to 0$ limit, $-C_1\rho_0 = 2C_2$, whence $C_2 = -C_1\rho_0/2$. So

$$u \sim C_1\left(\rho - \frac{\rho_0}{2}\rho^2\right) = C_1\left(\rho - n\rho^2\right) .$$

The rest of the analysis is the same as for the $l > 0$ case.

The truncated v series is the set of associated Laguerre polynomials $L_{-n-l-1}^{2l+1}\left(\frac{2r}{na}\right)$, and the full radial function $R(r) = u(r)/r$ is given by

$$R_{nl}(r) = N_{nl}(a)e^{-r/na}\left(\frac{2r}{na}\right)^l L_{-n-l-1}^{2l+1}\left(\frac{2r}{na}\right). \tag{2.59}$$

Figure 2.6 shows the effective potential including the centrifugal term for $l \geq 0$, the first three energy levels, and $u = rR$ wave functions. The full wave function is $\psi_{nlm}(r, \theta, \phi) = R_{nl}(r)Y_l^m(\theta, \phi)$.

We have used the definition $\rho \equiv \kappa r$. We have then used the standard definition of the Bohr radius a: $a \equiv 1/n\kappa$ to get the the function arguments r/na. (We have, as is customary, also absorbed powers of 2 into the arguments.) $N_{nl}(a)$ is the normalization constant.

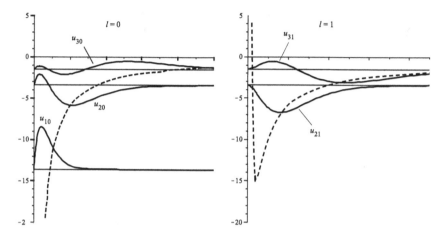

Fig. 2.6   The Coulomb potential for $l = 0$ and $l = 1$ and some energy levels and wave functions.

Problem 2.14. *The radial functions $R_{nl}(r)$ are dependent on l. For example,*

$$R_{2s}(r) = \frac{1}{2a} \sqrt{\frac{2}{a}} \left(1 - \frac{r}{2a}\right) e^{-r/2a} \; ; \quad R_{2p}(r) = \frac{1}{2a^2} \sqrt{\frac{1}{6a}} \; re^{-r/2a}$$

(a) *In spherical coordinates, the probability density $P(r)$ of finding the electron within dr of r is proportional to $r^2 R^2(r)$. Calculate the value of radius r at which you are most likely to find a 2s and a 2p electron.*

(b) *The labels s, p, etc. designate different values if angular momentum. Treating the values you have just found as the classical values of the radii of the electron orbits, explain why Bohr's quantization rule $L = n\hbar$ is inconsistent with classical mechanics. Hint: relate energy to angular momentum for a circular orbit. That is what Bohr did. Sommerfeld improved on Bohr's model by allowing for elliptical orbits.*

Problem 2.15. $R_{2s}(r) = \frac{1}{2a} \sqrt{\frac{2}{a}} \left(1 - \frac{r}{2a}\right) e^{-r/2a}$. *Recall that $u \equiv rR$. We found that for the s states ($l = 0$), $u \sim [\rho - n\rho^2]$ as $\rho \to 0$. Confirm this for u_{2s}. Hint: Be careful to keep all terms of the same order.*

This completes the analysis of our four examples using brute force solution of the Schrödinger equation. We hope that this chapter has been instructive, and tedious enough to whet your appetite for the more elegant SUSYQM approach.

Chapter 3

Operator Formalism in Quantum Mechanics

3.1 Operators and Representations

As we discussed in Chapter 1, the identification of the momentum with a differential operator led to the correspondence between the first term in the Schrödinger equation and the kinetic energy: $\mathbf{p} \equiv -i\hbar\nabla \rightarrow p^2/2m \equiv \frac{-\hbar^2}{2m}\nabla^2\Psi$, in our convention simply $-\nabla^2\Psi$. For simplicity we restrict ourselves to one dimension. p and x do not commute. The former is a differential operator, the latter simply a multiplier (or one might say a multiplicative operator). Explicitly,

$$p\,x\,\psi(x) = -i\frac{d}{dx}x\psi(x) = -ix\frac{d}{dx}\psi(x) - i\psi(x) , \qquad (3.1)$$

which we can write as $(xp - p\,x)\psi(x) = i\psi(x)$, where the term in parentheses is called the commutator of x and p, and written $[x, p]$. More elegantly, we write this as an operator equation: $[x, p] = i$; i.e. applying the commutator of x and p to any wave function reproduces an extra term: i times that wave function[1].

If we use the momentum space p rather than the coordinate space x, then the wave functions would become $\tilde{\phi}(p)$. In this case the momentum operator p would be a multiplier, and x would be the differential operator: $x \equiv id/dp$. But the operation $[x, p]$ on $\tilde{\phi}$ would still yield $i\tilde{\phi}$. This indicates that the operator formalism is more general. It does not depend on which representation we use for the wave function.

The hamiltonian in one-dimensional coordinate space is an operator comprising a differential operator $-\frac{d^2}{dx^2}$ and a multiplicative operator $V(x)$: $H = -\frac{d^2}{dx^2} + V(x)$.

---

[1]But one must take care to keep the function in while performing the operations, otherwise the extra term will tend to get lost.

3.2   Dirac Notation

Dirac invented a notation which not only simplified writing down the equations of quantum mechanics, but also was independent of representation. Its elements are brackets $\langle \, \rangle$, and he separated them into a left side $\langle \, |$ called a "bra" and a right side $| \, \rangle$ called a "ket." The kets represent the wave functions, the bras, their complex conjugates, and their bracket combination an integration of the two. The Dirac notation does not include the representation variable, in this case, x. It could just as well describe the integral in the momentum representation, or in any other. It contains only the physics. Thus $\langle m|n \rangle$ is the integral representing the overlap between states $|m\rangle$ and $|n\rangle$. Note that $\langle m|n \rangle$ is a number, just like the dot product in the space of real vectors. Since the spaces here are made up of complex vectors, the dot product must be generalized. It is called the inner product, with the property that $\langle m|n \rangle = \langle n|m \rangle^*$, in the coordinate representation, suppressing the limits of integration, $\int \psi_m^*(x)\psi_n(x)dx = (\int \psi_n^*(x)\psi_m(x)dx)^*$ An "inner product space" is one for which the inner product exists. This is a bit more subtle than it seems. For a finite inner product space of dimensionality N, there are N linearly independent vectors which span the space. For infinite-dimensional vector spaces, one requires that the inner product of the vectors and their limits exist. These are called Hilbert spaces, and it is in them that quantum mechanics operates.

The application of some operator A on a state $|n\rangle$ is designated $A|n\rangle \equiv |An\rangle$. The time-independent Schrödinger equation, for example, may be written as $|Hn\rangle = E_n|n\rangle$ irrespective of representation. However an inner product is written $\langle m|An \rangle$, to emphasize that the operator A acting on the ket $|n\rangle$ produces a new ket $|An\rangle$, whose overlap with the ket $|m\rangle$ is then computed.

The expectation value of an operator is written as $\langle n|An \rangle$, in the one-dimensional coordinate representation, $\int \psi_n^*(x)A\psi_n(x)dx$. More generally, $\langle m|An \rangle$, $\int \psi_m^*(x)A\psi_n(x)dx$ in the coordinate representation, represents a transition from an initial state $|n\rangle$ to a final state $|m\rangle$ due to the operator A. The only shortcoming of the Dirac notation is that it has no way to describe the probability of finding the particle in a given interval (a,b), $\int_a^b \psi^*(x)\psi(x)dx$. The bracket implies the integration over the entire domain.

Now how can we relate an operation on a ket to one on a bra? We define the "adjoint" or "hermitian conjugate" of operator A, $A^\dagger$:

$$\langle A^\dagger m|n \rangle \equiv \langle m|An \rangle. \tag{3.2}$$

That is, the adjoint of an operator acting on the bra gives the same result as the original operator acting on the ket. In the coordinate representation, this is $\int (A^\dagger \psi_m(x))^* \psi_n(x) dx = \int \psi_m^*(x) A\psi_n(x) dx$.

If a column vector represents a ket, its adjoint is the bra represented by a row vector with components that are the complex conjugates of those in the column vector:

$$|n\rangle \equiv \begin{pmatrix} n_1 \\ n_2 \\ \vdots \end{pmatrix} \qquad \langle n| \equiv \begin{pmatrix} n_1^* & n_2^* \cdots \end{pmatrix} .$$

In two dimensions, $A \equiv \begin{pmatrix} a_{11} & a_{12} \\ a_{21} & a_{22} \end{pmatrix}$, $A^\dagger \equiv \begin{pmatrix} a_{11}^* & a_{21}^* \\ a_{12}^* & a_{22}^* \end{pmatrix}$, $|n\rangle \equiv \begin{pmatrix} n_1 \\ n_2 \end{pmatrix}$, and $\langle m| \equiv \begin{pmatrix} m_1^* & m_2^* \end{pmatrix}$, the adjoint of $|m\rangle \equiv \begin{pmatrix} m_1 \\ m_2 \end{pmatrix}$.

Problem 3.1. *In this two-dimensional matrix formulation, show that* $\langle A^\dagger m|n\rangle = \langle m|An\rangle$.

The rule for matrix multiplication $U = ST$ is $U_{ik} = \Sigma_j S_{ij} T_{jk}$. The hermitian conjugate of this product is the product taken in reverse order of the hermitian conjugates of the component matrices: $U^\dagger = T^\dagger S^\dagger$.

3.2.1 *Hermitian operators*

We have seen several examples of physical quantities: position, momentum, energy, being represented by operators. In quantum mechanics all physical observables are represented by operators. Measurements of a physical observable yield the eigenvalues of its associated operator. Hence the eigenvalues of quantum operators must be real. This constrains the form of the operators to be self-adjoint: $A^\dagger = A$. Self-adjoint operators are also known as hermitian operators.

For states $|n\rangle$ which are eigenvectors[2] of a hermitian operator A: $A|n\rangle = \lambda_n|n\rangle$, we can deduce that the eigenvalues λ_n must be real numbers : $\langle n|An\rangle = \langle n|\lambda_n|n\rangle = \lambda_n\langle n|n\rangle$. For a hermitian operator, $\langle n|An\rangle = \langle An|n\rangle = \langle \lambda_n n|n\rangle = \lambda_n^*\langle n|n\rangle$. Therefore $\lambda_n^* = \lambda_n$.

Problem 3.2. *In the one-dimensional coordinate representation, $p = -i\, d/dx$. Assuming that the wave function and its complex conjugate vanish*

---

[2] We shall use the terms "eigenvector" and "eigenstate" interchangeably.

at the boundaries, show using integration by parts that $p = i\,d/dx$ is hermitian. Is d/dx hermitian?

Problem 3.3.

(a) Prove that the sum of two Hermitian operators is Hermitian.

(b) Prove that if operator A is Hermitian, then so is A^n.

A particle can be in an eigenstate of some operator. But it can also be in a mixture: a linear combination of eigenstates. In this case we are interested in two things. The first is the expectation value of a particular operator A; i.e, the average value of the associated observable after many (technically infinitely many) measurements. But since any measurement can only return an eigenvalue of the operator, we are also interested in the probability that a particular eigenvalue is returned. We would like to employ the eigenvectors of the operator as the basis set. We will require them to be orthonormal: $\langle m|n\rangle = \delta_{mn}$. If they are not, we can make them so by procedures such as Gram-Schmidt orthogonalization [3].

Once the eigenvectors constitute an orthonormal set, we can show the following:

If a mixed state is given by $|\psi\rangle = \sum_n c_n |n\rangle$ where $A|n\rangle = \lambda_n |n\rangle$ then the probability of obtaining the result λ_n is $|\langle\psi|n\rangle|^2 = |c_n|^2$. We know that any state $|\psi\rangle$ must be normalized: the probability of finding the particle in that state somewhere in the domain must be 1. To exploit the orthonormality of the eigenstates, we can without loss of generality, write the mixed state as $|\psi\rangle = \sum_m c_m |m\rangle$.

Then $1 = \langle\psi|\psi\rangle = \sum_m c_m^* \langle m| \sum_n c_n |n\rangle = \sum_{mn} c_m^* c_n \langle m|n\rangle = \sum_{mn} c_m^* c_n \delta_{mn} = \sum_n |c_n|^2$.

Since the total probability of finding the particle is the sum of the probabilities for the particle to be in each of the eigenstates, this implies that $|c_n|^2$ represents the probability that a measurement of $|\psi\rangle$ will find it in state $|n\rangle$; i.e., the result of such a measurement will be the eigenvalue λ_n.

We can also think of this as follows: Since $|\psi\rangle = \sum_m c_m |m\rangle$, $\langle n|\psi\rangle = \langle n| \sum_m c_m |m\rangle = \sum_m c_m \delta_{mn} = c_n$. In words, the probability is the square of the overlap.

---

[3] The street grid of Chicago is orthonormal: there are eight blocks per mile north-south or east-west. New York's grid is orthogonal but needs to be normalized: there are 20 blocks per mile north-south, but 8 blocks per mile east-west. Paris, on the other hand is crying out for the full Gram-Schmidt procedure.

Problem 3.4. *In Chapter 2 we showed that the wave functions and energies for an infinite square well of length π were given by $|n\rangle = \sqrt{\frac{2}{\pi}} \sin nx$, $E_n = n^2$.*

(a) Show that the eigenfunctions constitute an orthonormal set.

A state of a particle in the well is in a mixture of the two lowest states: $\psi = \frac{1}{2}|1\rangle + \frac{\sqrt{3}}{2}|2\rangle$.

(b) Check that ψ is normalized.

(c) What is the expectation value of the energy? Does this correspond to any single measurement?

(d) What is the probability that a measurement of the energy returns the value 4?

Physical observables are of course not only the expectation values of operators for particles in a given state. They can also describe measurable quantities for particles changing state. For example, the "transition probability" is a measure of the likelihood of an electron jumping from one orbit to another, in Bohr model terminology. It is proportional to $|\langle m|An\rangle|^2$.

3.3 Constants of Motion in Quantum Mechanics

Constants of motion are conserved quantities. What does a conservation law mean in quantum mechanics? In classical mechanics the measurable quantities are dynamical variables such as position, momentum, angular momentum, and energy. In quantum mechanics, these dynamical quantities become operators. The best one can do is to obtain the expectation value associated with an operator. That is, to say the physical observable associated with some operator A is "conserved" means that $d\langle A\rangle/dt = 0$, in which case we call the eigenvalues of A "good quantum numbers." We have

$$\frac{d\langle A\rangle}{dt} = \frac{d\langle \Psi|A\Psi\rangle}{dt} = \left\langle \frac{\partial \Psi}{\partial t} \middle| A\Psi \right\rangle + \left\langle \Psi \middle| \frac{\partial A}{\partial t}\Psi \right\rangle + \left\langle \Psi \middle| A\frac{\partial \Psi}{\partial t} \right\rangle.$$

The time dependent Schrödinger equation relates such conservation to the hamiltonian, as follows:

$$|H\Psi\rangle = \left| i\frac{\partial \Psi}{\partial t} \right\rangle$$

and H is hermitian. So

$$\frac{d\langle A\rangle}{dt} = i\,\langle H\Psi|A\Psi\rangle + \left\langle\frac{\partial A}{\partial t}\right\rangle - i\,\langle\Psi|AH\Psi\rangle.$$

Rearranging,

$$\frac{d\langle A\rangle}{dt} = i\,\langle[H,A]\rangle + \left\langle\frac{\partial A}{\partial t}\right\rangle. \tag{3.3}$$

If A has an explicit time dependence, for example, a time-dependent potential $A \equiv V(t)$, then the full time dependence is the sum of both terms on the right of Eq. 3.3. For A not explicitly time-dependent, $\langle\frac{\partial A}{\partial t}\rangle = 0$, and the time dependence of $\langle A\rangle$ is determined entirely from its commutator with H.

A simple example is the identity operator **1**. Since it commutes with everything, $\frac{d\langle 1\rangle}{dt} = 0$. But this is just $d\langle\Psi|\Psi\rangle/dt$. This is a statement of conservation of normalization; viz., the probability of finding a particle somewhere is 1 at all times. Another example is $A \equiv x$. Ignoring our convention, putting back in $2m$,

$$\frac{d\langle x\rangle}{dt} = i\,\langle[H,x]\rangle = i\left\langle\left[\frac{p^2}{2m}+V(x),x\right]\right\rangle = i\left\langle\left[\frac{p^2}{2m},x\right]\right\rangle \tag{3.4}$$

where we have used that $[V(x),x] = 0$.

Problem 3.5. *Using the Taylor expansion of $V(x)$ prove that $[V(x),x] = 0$.*

But $[p^2,x] = ppx - xpp$, where $xp - px = i$. Then $ppx = p\,(xp - i)$ and $xpp = (xp)p = (px + i)p$. Canceling the terms $p\,xp$, $[H,x] = -2ip/2m$. Therefore

$$m\frac{d\langle x\rangle}{dt} = \langle p\rangle, \tag{3.5}$$

which is the quantum mechanical definition of momentum.

Problem 3.6. *(Ehrenfest's theorem) Show that*

$$\frac{d\langle p\rangle}{dt} = \left\langle-\frac{dV}{dx}\right\rangle.$$

Thus, we see the importance of the hamiltonian: if $[H,A] = 0$ then $\langle A\rangle$ is a constant of the motion. In retrospect, this comes as no surprise. The time dependent Schrödinger equation links H with the time evolution of the wave function, from which the time evolution of the expectation value of an operator is obtained.

Another way of considering the consequences of commutation with the hamiltonian is to recognize that the eigenstates of the hamiltonian are the same as those for the operator(s) with which it commutes. Thus, we write the wave functions for the hydrogen atom (ignoring spin) as $\psi_{n\ell m}$, indicating that the total angular momentum and its projection commute with the hamiltonian.

Let us prove a general result: *Commuting operators have the same eigenvectors.*

Proof: Suppose $[A, B] = 0$ and $|An\rangle = \lambda_n|n\rangle$. Then, since $BA = AB$, $AB|n\rangle = B\lambda_n|n\rangle = \lambda_n|Bn\rangle$. This says that $|Bn\rangle$ is an eigenstate of A with the same eigenvalue as $|n\rangle$ itself. This can only be the case if $|Bn\rangle$ is parallel to $|n\rangle$ itself; i.e. $|Bn\rangle = \mu_n|n\rangle$. Thus $|n\rangle$ is an eigenstate of B with eigenvalue μ_n.

While this is a general result, its application to the hamiltonian is the very definition of good quantum number. It provides the basis for the convenient vector diagrams of, for example, the orbital angular momentum and the spin precessing around the total angular momentum in the description of spin-orbit coupling in Modern Physics courses. It tells us which quantum numbers become "bad" as a hamiltonian is changed, for example as we go from the Schrödinger to the Dirac equation. It has shown that some so-called "accidental degeneracies" are not accidental at all, by the discovery of hitherto unknown operators commuting with the hamiltonians, and has provided new constants of the motion. Operator algebra has enabled us to discover these and other underlying physics, without the need for choosing any particular representation. It is indispensable for SUSYQM.

Chapter 4

Algebraic Solution for the Harmonic Oscillator

In Chapter 2 we have seen that rather tedious computations yield the solutions (eigenenergies and eigenfunctions) for some of the elementary systems we looked at. However, there is a different path to these results. The method we are about to show for the harmonic oscillator case doesn't involve solving the Schrödinger equation, while obtaining the same results. The way is provided by an algebraic method often attributed to Dirac [1]. At the core of this algebraic method is finding a set of specially crafted operators A^- and A^+ that will transform one eigenfunction of the hamiltonian into another eigenfunction with a different energy. In this way, starting from any eigenfunction ψ_n, one can obtain all the other eigenfunctions by successive application of A^- or A^+.

To begin, let us rewrite the Schrödinger equation for the harmonic oscillator, Eq. (2.21), as the eigenvalue problem for hamiltonian operator $H\psi(x) = E\psi(x)$, where

$$H = -\frac{d^2}{dx^2} + \frac{1}{4}\,\omega^2\,x^2 \ . \tag{4.1}$$

Next, we define two operators

$$A^- \equiv \frac{1}{\sqrt{\omega}}\left(\frac{d}{dx} + \frac{1}{2}\,\omega\,x\right) \ , \quad A^+ \equiv \frac{1}{\sqrt{\omega}}\left(-\frac{d}{dx} + \frac{1}{2}\,\omega\,x\right) \ . \tag{4.2}$$

which are known respectively as *annihilation* and *creation* operators. (The reason for these names will become clear shortly.) We can immediately check, by operating on an arbitrary function ψ with them, that $A^-\,(A^+\psi) - A^+\,(A^-\psi) = \psi$. This property of the operators is usually written as a commutation relation:

$$A^-A^+ - A^+A^- \equiv [A^-, A^+] = \mathbf{1} \ , \tag{4.3}$$

[1]But see footnote 1 in Chapter 1.

37

where **1** is the identity operator. To simplify notation, it is customary to simply write the number 1. The product operator $N \equiv A^+ A^-$ is Hermitian.

Problem 4.1. *Using the results of Problem 3.2, prove N is Hermitian.*

Consequently, all its eigenvalues are real. Let us denote by $|n\rangle$ the eigenvectors of N. Then

$$N|n\rangle = n|n\rangle . \qquad (4.4)$$

Furthermore, we can show that N is a linear function of the hamiltonian.

$$
\begin{aligned}
N\psi &= A^+ A^- \psi \\
&= \frac{1}{\omega}\left(-\frac{d}{dx} + \frac{1}{2}\omega x\right)\left(-\frac{d}{dx} + \frac{1}{2}\omega x\right)\psi \\
&= \frac{H}{\omega}\psi - \frac{1}{2}\psi .
\end{aligned}
\qquad (4.5)
$$

Therefore the two operators N and H commute, and consequently, they have a common set of eigenvectors. Hence $|n\rangle$ is an energy eigenvector as well. From (4.5) we have

$$H = \omega\left(N + \frac{1}{2}\right) , \qquad (4.6)$$

and therefore,

$$H|n\rangle = \left(n + \frac{1}{2}\right)\omega|n\rangle , \qquad (4.7)$$

which means that the eigenenergies of the harmonic oscillator are given by

$$E_n = \left(n + \frac{1}{2}\right)\omega . \qquad (4.8)$$

Problem 4.2. *Repeat the above analysis for the operator $M \equiv A^- A^+$. How are its eigenvalues m related to n?*

At this stage, we should be careful when comparing Eq. (2.31) with Eq. (4.8). So far, all we know in Eq. (4.8) is that n is a non-negative real number. Indeed, from the positivity of the norm of $(A^-|n\rangle)$ one gets

$$n = \langle n|N|n\rangle = \left(\langle n|A^+\right)\left(A^-|n\rangle\right) \geq 0 . \qquad (4.9)$$

To find the nature of the positive number n, we will explore some properties of A^- and A^+ operators. Using the commutation relation (4.3) we note that

$$[N, A^+] = [A^+ A^-, A^+] = A^+[A^-, A^+] + [A^+, A^+]A^- = A^+ , \qquad (4.10)$$

and, similarly

$$[N, A^-] = -A^- . \tag{4.11}$$

We can now prove that if $|n\rangle$ is an eigenvector of N, then $A^+|n\rangle$ and $A^-|n\rangle$ are also eigenvectors of N. Indeed,

$$N A^+|n\rangle = ([N, A^+] + A^+ N)|n\rangle = (n+1) A^+|n\rangle , \tag{4.12}$$

$$N A^-|n\rangle = ([N, A^-] + A^- N)|n\rangle = (n-1) A^-|n\rangle . \tag{4.13}$$

We observe that $A^+|n\rangle$ is an eigenvector of N with eigenvalue increased by 1, and we write accordingly

$$A^+|n\rangle = c_+ |n+1\rangle , \tag{4.14}$$

where c_+ is a constant to be determined using the normalization condition $\langle n+1 | n+1 \rangle = 1$. We have

$$\langle n | A^- A^+ | n\rangle = |c_+|^2 . \tag{4.15}$$

But $A^- A^+ = 1 + A^+ A^- = 1 + N$, and therefore the left hand side of Eq. (4.15) is equal to $n+1$. Hence, $|c_+|^2 = n+1$, and Eq. (4.14) becomes

$$A^+|n\rangle = \sqrt{n+1}\,|n+1\rangle . \tag{4.16}$$

Similarly,

$$A^-|n\rangle = \sqrt{n}\,|n-1\rangle . \tag{4.17}$$

Now, we can see why A^+ (A^-) is called a *creation* (*annihilation*) operator. Acting on the energy eigenvector $|n\rangle$, this operator yields another energy eigenvector $|n+1\rangle$ ($|n-1\rangle$) whose corresponding eigenenergy is increased (decreased) by one quantum of energy ω.

We are now ready to completely identify the positive real number n. Let us start from an arbitrary state $|n\rangle$ and successively apply the annihilation operator A^-. We obtain

$$A^-|n\rangle = \sqrt{n}\,|n-1\rangle ,$$
$$(A^-)^2|n\rangle = \sqrt{n(n-1)}\,|n-2\rangle ,$$
$$\vdots$$
$$(A^-)^k|n\rangle = \sqrt{n(n-1)\cdots(n-k+1)}\,|n-k\rangle , \tag{4.18}$$
$$\vdots$$

If n is an arbitrary *non-integer* positive real number, then after applying the annihilation operator A^- an integral number of times k, we must overshoot

0 and arrive at an eigenvector $|n - k\rangle$ whose eigenvalue $n - k$ is negative. This contradicts Eq. (4.9). Therefore, n must be a positive integer, and consequently, the sequence (4.18) terminates at the eigenvector of N with the lowest possible eigenvalue. We call this eigenvector $|0\rangle$. If we go any further, the sequence terminates, since from Eq. (4.18), the eigenvalue of $|0\rangle$ is 0.

$$A^- |0\rangle = 0 . \tag{4.19}$$

The vector $|n - k\rangle = |0\rangle$ is the ground state vector, with the lowest energy $E_0 = \omega/2$. Now, we can finally conclude that the numbers n in Eq. (4.8) are non-negative integers. Hence, we recover the same result as in Eq. (2.31), but this time via the less cumbersome algebraic method.

Starting from the ground state $|0\rangle$ we can successively apply the creation operator A^+ to obtain any state $|n\rangle$. Using (4.16), we obtain

$$|1\rangle = A^+ |0\rangle ,$$

$$|2\rangle = \frac{A^+}{\sqrt{2}} |1\rangle = \frac{(A^+)^2}{\sqrt{2}} |0\rangle ,$$

$$|3\rangle = \frac{A^+}{\sqrt{3}} |2\rangle = \frac{(A^+)^3}{\sqrt{3!}} |0\rangle ,$$

$$\vdots$$

$$|n\rangle = \frac{A^+}{\sqrt{n}} |n - 1\rangle = \frac{(A^+)^n}{\sqrt{n!}} |0\rangle . \tag{4.20}$$

So far we have used the abstract representation of the eigenvectors of the hamiltonian of the system. We make contact with the familiar coordinate representation, where the energy eigenvectors become simply the well-known wave functions via $|n\rangle \to \psi_n(x)$. We can use the "differential realization" of the creation operator A^+ from Eq. (4.2) in conjunction with Eq. (4.20) to find all $\psi_n(x)$ if we know the ground state $\psi_0(x)$. To find $\psi_0(x)$, we rewrite Eq. (4.19) as a simple differential equation

$$\frac{1}{\sqrt{\omega}} \left(\frac{d}{dx} + \frac{1}{2}\omega x \right) \psi_0(x) = 0 . \tag{4.21}$$

using the differential realization of A^- from (4.2). Solving this first order homogeneous differential equation, one gets

$$\psi_0(x) = \left(\frac{\omega}{2\pi} \right)^{\frac{1}{4}} e^{-\frac{1}{4}\omega x^2} , \tag{4.22}$$

where the integration constant was determined from the normalization condition $\int_{-\infty}^{+\infty} |\psi_0(x)|^2 = 1$.

Problem 4.3. *Obtain Eq. (4.22)from Eq. (4.21).*

Substituting Eq. (4.21) into Eq. (4.20) we obtain

$$\psi_1(x) = \frac{1}{\sqrt{\omega}} \left(-\frac{d}{dx} + \frac{1}{2}\omega x \right) \left(\frac{\omega}{2\pi} \right)^{\frac{1}{4}} e^{-\frac{1}{4}\omega x^2}$$

$$\psi_2(x) = \frac{1}{\sqrt{2!}} \left(\frac{1}{\sqrt{\omega}} \right)^2 \left(-\frac{d}{dx} + \frac{1}{2}\omega x \right)^2 \left(\frac{\omega}{2\pi} \right)^{\frac{1}{4}} e^{-\frac{1}{4}\omega x^2}$$

$$\vdots$$

$$\psi_n(x) = \frac{1}{\sqrt{n!}} \left(\frac{1}{\sqrt{\omega}} \right)^n \left(-\frac{d}{dx} + \frac{1}{2}\omega x \right)^n \left(\frac{\omega}{2\pi} \right)^{\frac{1}{4}} e^{-\frac{1}{4}\omega x^2} \qquad (4.23)$$

which can be brought into the same form as Eq. (2.36) using Rodrigues formula, for the Hermite polynomials.

We now see that Eqs. (2.38) and (2.39) are nothing but the reflection of the action of the annihilation and creation operators on the wave function $\psi_n(x)$. Because of their special action on the energy eigenfunctions, the operators A^- and A^+ are also known as *ladder operators*. We will see in the next chapter that many potentials allow for a generalization of the algebraic method introduced here. The generalization of this method to other potentials is SUSYQM.

Chapter 5

Supersymmetric Quantum Mechanics

In Chapter 1, we gave a brief introduction to SUSYQM. In this chapter we will use it to solve several problems. We will introduce it more formally in Chapter 7.

5.1 The Concept of Superpotential

As we saw in Chapter 2, interactions in quantum mechanics are described by a potential function $V(x, a)$, where x is the coordinate variable and the parameter a describes the "strength" of the interaction. For example, in the harmonic oscillator, the potential is $V(x, \omega) = \frac{1}{4}\omega^2 x^2$. Here ω measures the curvature of the potential function. The energy of the system $E_n = (n + \frac{1}{2})\omega$ depends on this parameter. In the Coulomb problem, the effective potential depends on three arguments: $V(x, e, l) = \left[-\frac{e^2}{r} + \frac{l(l+1)}{r^2}\right]$. Interactions depend upon charges e and orbital angular momentum l. It is in this sense that we say that these parameters, which we generically denote by a, define the "strength" of the potential.

Our objective is to solve the Schrödinger equation to determine the complete set of eigenvalues E_n and eigenfunctions $\psi_n(x, a)$ for a given hamiltonian H:

$$H \psi(x, a) \equiv \left[-\frac{d^2}{dx^2} + V(x, a)\right] \psi(x, a) = E \psi(x, a) \ . \qquad (5.1)$$

In Chapter 2, we considered several examples of eigenvalue equations and we solved them with varying levels of effort. In Chapter 4 we solved one of those examples; namely, the harmonic oscillator, using a scheme based on algebra, rather than solving a differential equation. In this chapter we will show that the algebraic method used for the harmonic oscillator can be extended to other quantum mechanical systems. However, it is worth noting that there is a frequently overlooked benefit associated with the traditional method. In Chapter 2, after rather tedious calculations, we not only found all eigenvalues of the hamiltonian, we also described all eigenfunctions in closed form as we did in Eq. (2.36) for the harmonic oscillator. SUSYQM takes a different approach. It provides a simpler method of deriving the energy eigenvalues. However, it only determines the eigenfunctions recursively.

In the case of the harmonic oscillator, we were able to factorize the hamiltonian $H = \left(-\frac{d^2}{dx^2} + \frac{1}{4}\omega^2 x^2\right)$ into a product of two first order operators and an additive constant: $A^+ A^- + \frac{1}{2}\omega$, where $A^- = \frac{d}{dx} + \frac{1}{2}\omega x$ and $A^+ = -\frac{d}{dx} + \frac{1}{2}\omega x$. We will generalize this method for an arbitrary potential $V(x,a)$. In order to factorize the hamiltonian, we will first need to express the potential $V(x,a)$ differently. As $\frac{1}{2}\omega x$ in $A^\pm$ was related to the potential $\frac{1}{4}\omega^2 x^2$ for the harmonic oscillator, we define a real function $W(x,a)$, that can be related to $V(x,a)$ in SUSYQM formalism. This progenitor $W(x,a)$ of the potential function is known as the superpotential. As we will see shortly, a superpotential $W(x,a)$ generates two potentials $V_\pm(x,a)$ that we call partners. We introduce two operators $A^+ = -\frac{d}{dx} + W(x,a)$ and $A^- = \frac{d}{dx} + W(x,a)$. The action of their product $A^+ A^-$ on any general function $\psi(x,a)$ is equivalent to the action of the operator $-\frac{d^2}{dx^2} + \left[W^2(x,a) - \frac{dW(x,a)}{dx}\right]$. I.e., the Schrödinger equation for a quantum mechanical system described by a potential $V(x,a) = W^2(x,a) - \frac{dW(x,a)}{dx}$ can be factorized as

$$A^+ A^- \psi(x,a) \equiv \left(-\frac{d}{dx} + W(x,a)\right)\left(\frac{d}{dx} + W(x,a)\right)\psi(x,a)$$

$$= \left[-\frac{d^2}{dx^2} + W^2(x,a) - \frac{dW(x,a)}{dx}\right]\psi(x,a) . \quad (5.2)$$

If we interchange the operators A^- and A^+, we obtain

$$A^- A^+ \psi(x,a) \equiv \left(\frac{d}{dx} + W(x,a)\right)\left(-\frac{d}{dx} + W(x,a)\right)\psi(x,a)$$

$$= \left[-\frac{d^2}{dx^2} + W^2(x,a) + \frac{dW(x,a)}{dx}\right]\psi(x,a) , \quad (5.3)$$

which is again a Schrödinger equation, but this time for a different potential $W^2(x,a) + \frac{dW(x,a)}{dx}$. Thus, a superpotential $W(x,a)$ helps us factorize two hamiltonians, which we will denote by $H_\pm \equiv -\frac{d^2}{dx^2} + V_\pm(x,a)$, associated with potentials

$$V_\pm(x,a) = W^2(x,a) \pm \frac{dW(x,a)}{dx} . \tag{5.4}$$

We will denote eigenvalues and eigenfunctions of the hamiltonians $H_\pm$ by $E_n^\pm$ and $\psi_n^\pm(x,a)$. We can make a few interesting observations here.

(1) Suppose $\psi_0^-(x,a)$, the solution of $A^- \psi_0^-(x,a) = 0$ is a normalizable function. Then $H_- \psi_0^-(x,a) \equiv A^+ A^- \psi_0^-(x,a) = 0$, ;i.e., $\psi_0^-(x,a)$ is an eigenstate of H_- with an eigenvalue zero.

(2) Since A^+ is the adjoint or hermitian conjugate[1] of A^-, from $H_- \psi_n^-(x,a) = A^+ A^- \psi_n^-(x,a) = E_n^- \psi_n^-(x,a)$, we obtain

$$E_n^- = \frac{\int \psi_n^{-*} A^+ A^- \psi_n^- \, dx}{\int \psi_n^{-*} \psi_n^- \, dx} = \frac{\int |A^- \psi_n^-|^2 \, dx}{\int |\psi_n^-|^2 \, dx} \geq 0 . \tag{5.5}$$

Thus, the eigenvalues of the operator $H_- \equiv A^+ A^-$ are greater than or equal to zero. In such cases, we call the operator $A^+ A^-$ a "semi-positive definite" operator.

Problem 5.1. *Show that the ground state eigenfunction of $\psi_0^-(x,a)$ can be written as $\psi_0^-(x,a) \sim e^{-\int_{x_0}^x W(x,a)\,dx}$.*

Analogously, if we had instead started with the solution of $A^+ \psi_0^+(x,a) = 0$, we would get

$$H_+ \psi_0^+(x,a) \equiv A^- A^+ \psi_0^+(x,a) = 0 ; \tag{5.6}$$

and thus the zero energy ground state eigenfunction of H_+ is given by the solution of

$$\left(-\frac{d}{dx} + W(x,a) \right) \psi^+(x,a) = 0 ,$$

i.e.

$$\psi_0^+(x,a) \sim e^{\int_{x_0}^x W(x,a)\,dx} \sim \frac{1}{\psi_0^-(x,a)} .$$

From this reciprocal relationship between $\psi_0^-(x,a)$ and $\psi_0^+(x,a)$, we see that they cannot be normalized simultaneously. This leads to three different possibilities:

---

[1]These two operators are hermitian conjugates of each other, i.e., for any two functions $\eta(x)$ and $\lambda(x)$, we have $\int [A^+\eta(x)]^* \lambda(x)\, dx = \int \eta(x)^* A^- [\lambda(x)]\, dx$ and $\int [A^-\eta(x)]^* \lambda(x)\, dx = \int \eta(x)^* A^+ [\lambda(x)]\, dx$.

(1) $\psi_0^-(x,a)$ is normalizable and $\psi_0^+(x,a)$ is not;
(2) $\psi_0^+(x,a)$ is normalizable and $\psi_0^-(x,a)$ is not;
(3) neither $\psi_0^-(x,a)$ nor $\psi_0^+(x,a)$ is normalizable.

5.1.1 *Broken and Unbroken Supersymmetry*

Let us consider the above three possibilities. In the first two cases, the ground state energies of H_- and H_+ are respectively zero. In the third case neither of the two hamiltonians has a zero-energy bound state. When either of the two partner hamiltonians has a zero energy bound state, the system is said to possess *unbroken* supersymmetry, and when neither holds a zero energy bound state the supersymmetry is said to be *spontaneously broken*. We will explain the reasons behind this nomenclature in Chapter 7.

Let us explore which properties of $W(x,a)$ are important in determining whether supersymmetry is broken or unbroken. Normalizability of $\psi_0^-(x,a)$ requires that the wave function vanish as $x \to \pm\infty$. This implies that $\psi_0^-(\pm\infty, a) \sim e^{-\int_{x_0}^{\pm\infty} W(x,a)\,dx} = 0$, and hence $\int_{x_0}^{\pm\infty} W(x,a)\,dx = \infty$. We consider the integrals separately. $\int_{x_0}^{+\infty} W(x,a)\,dx$ is the area under the curve from x_0 to ∞ and is $+\infty$. This implies that the superpotential $W(x,a)$ must have a positive value at $+\infty$. However, $\int_{x_0}^{-\infty} W(x,a)\,dx = -\int_{-\infty}^{x_0} W(x,a)\,dx = +\infty$ implies that the area under the curve from $-\infty$ to x_0 must be $-\infty$. Consequently, $W(x,a)$ must have a negative value at $-\infty$. Hence, $W(x,a)$ must cross the x-axis. This is illustrated in Figure (5.1).

While any odd function $W(x)$ with the property $\int_{x_0}^{\pm\infty} W(x,a)\,dx = \infty$ will lead to an unbroken supersymmetry, odd parity is not necessary. For example, while the odd function $W(x,A) = A\tanh x$ leads to a system with unbroken SUSY, so does the function $W(x,A,B) = A\tanh x + B$ with $A > B > 0$ which has no particular parity. In short, a superpotential that has the same sign at both ends of the domain will generate a system with broken supersymmetry, and one that has opposite sign will lead to unbroken supersymmetry.

Note that for cases that have finite domain, $x = -\infty$ and $x = +\infty$ are to be replaced by x_L and x_R, the left and right boundary points, respectively. Normalization of $\psi_0^\pm$ would then require that $W(x,a)$ diverge at both the finite boundary points. From here onward we will assume, without loss of generality, that for all cases of unbroken SUSY the superpotential is negative at the left and positive at the right boundary.

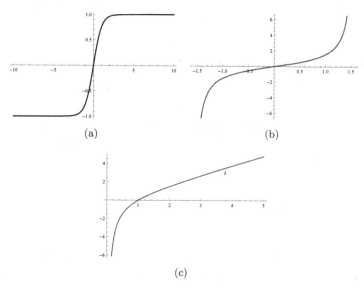

Fig. 5.1   Examples of superpotentials $W(x)$ that lead to unbroken supersymmetry for (a) an infinite domain, (b) a finite domain and (c) a semi-infinite domain.

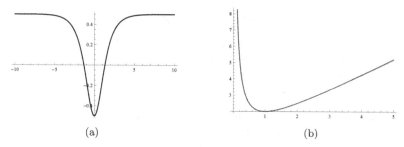

Fig. 5.2   Examples of superpotentials $W(x)$ that lead to broken supersymmetry for (a) an infinite domain and (b) a semi-infinite domain.

Problem 5.2. *For the following three cases, compute ground state wave functions and show that they vanish at both end points of their domains:*

(1) Finite domain: $W(x, \alpha) = \alpha \tan x$ *with* $\alpha > 0$;

(2) Semi-infinite domain: $W(x, \alpha, \beta) = \alpha - \beta \coth x$ *with* $\alpha > \beta > 0$; *and*

(3) Infinite domain: $W(x, \gamma) = \gamma \tanh x$ *with* $\gamma > 0$.

Problem 5.3. *The superpotential $W(r, \omega, l)$ for the three-dimensional harmonic oscillator given by $W(r, \omega, l) = \omega r - \frac{l}{r}$,*

(1) plot $W(r, 10, -1)$ as a function of r;
(2) show that the function $\exp\left(-\int_0^\infty W(r, 10, -1)\, dr\right)$ does not converge at least one boundary point.

Since the two hamiltonians H_- and H_+ originate from one common superpotential, we shall soon see that they share many common properties. They are called partner hamiltonians and their potentials are known as partner potentials.

5.2   Isospectrality

One of the interesting relationships between two partner hamiltonians is known as isospectrality. That is, except possibly for the ground state of H_-, both hamiltonians have exactly the same set of eigenvalues and corresponding eigenfunctions, related by an operation of A^- or A^+. Let $\psi_n^\pm$ denote the eigenfunctions of $H_\pm$ that correspond to eigenvalues $E_n^\pm$; $n = 1, 2, \ldots$

$$H_+ \left(A^- \psi_n^-\right) = A^- A^+ \left(A^- \psi_n^-\right) = A^- \left(A^+ A^- \psi_n^-\right)$$
$$= A^- \left(H_- \psi_n^-\right) = E_n^- \left(A^- \psi_n^-\right) . \qquad (5.7)$$

Thus, for every eigenstate ψ_n^- of H_-, the function $A^- \psi_n^-$ is an eigenstate of H_+, and the corresponding eigenvalue is E_n^-. All excited states ψ_n^- have a one-to-one correspondence with ψ_{n-1}^+:

$$\psi_{n-1}^+ = c A^- \psi_n^- \quad \text{and} \quad E_{n-1}^+ = E_n^-, \qquad \text{where } n = 1, 2, \ldots . \quad (5.8)$$

Why in Eq. (5.8) are we ignoring the eigenfunction with $n = 0$? If the supersymmetry is unbroken, the ground state of H_- obeys $A^- \psi_0^- = 0$ and it does not connect to any state of H_+. On the other hand, if supersymmetry is broken, $A^- \psi_0^- = 0$ does not have a normalizable solution, we would have a strict isospectrality between two hamiltonians; i.e., $E_n^+ = E_n^-$ for all values of n. Thus in each case, if we knew the eigenvalues and eigenfunctions of either of the two partner hamiltonians, we could determine the spectrum of the other.

Problem 5.4. *Determine the normalization constants to show that normalized eigenfunctions $\psi_n^\pm$ are related by*

$$\psi_n^+ = \frac{1}{\sqrt{E_{n+1}^-}} A^- \psi_{n+1}^- ; \quad \text{and} \quad \psi_n^- = \frac{1}{\sqrt{E_{n-1}^+}} A^+ \psi_{n-1}^+ . \qquad (5.9)$$

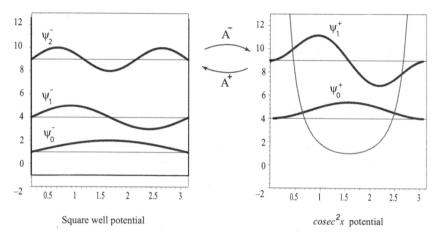

Fig. 5.3 The infinite square well and its $\cosec^2 x$ partner potential, together with a few energy levels and the corresponding eigenfunctions. Since this is a case of unbroken supersymmetry, note that operators A^- and A^+ respectively destroy and create nodes of wave functions they act upon.

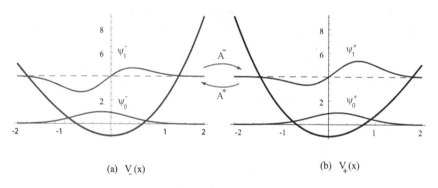

Fig. 5.4 A schematic diagram for two partner potentials for a broken SUSY case. Note that operators A^- and A^+ do not alter the nodal structure of eigenfunctions upon which they act. In addition, the ground state energies $E_0^{(-)}$ and $E_0^{(+)}$ are greater than zero.

Here is an example specially selected to exhibit the efficacy of this formalism. A judiciously chosen superpotential in this example will yield two partner potentials: one the well known infinite well potential and the other the much less familiar $\cosec^2 x$ (the trigonometric Pöschl-Teller or Pöschl-Teller I) potential. We determined spectra for both of these two potentials in Chapter 2, one very easily and the other with considerable difficulty, and found that they were the same except for the ground state for domain $(0, \pi)$. In this chapter, we will solve the second from the first, using SUSYQM formalism. We begin by showing that they are superpotential partners.

5.3   Pöschl-Teller Potential

Consider a superpotential $W(x) = -b \cot x$ with the parameter $b > 0$, and domain $(0, \pi)$. The supersymmetric partner potentials are:

$$V_-(x, b) = W^2(x) - \frac{dW}{dx} = b(b-1)\operatorname{cosec}^2 x - b^2$$

and   (5.10)

$$V_+(x, b) = W^2(x) + \frac{dW}{dx} = b(b+1)\operatorname{cosec}^2 x - b^2 \ .$$

Now for the special case of $b = 1$, the potential $V_-(x, 1)$ is a trivial constant function $-b^2 = -1$; i.e., just an infinite one-dimensional square well potential whose bottom is set to -1 and whose boundaries are at 0 and π, while its partner potential $V_+(x, 1)$ is given by $(2\operatorname{cosec}^2 x - 1)$. Thus, $V_-(x, 1)$ and $V_+(x, 1)$, two very different potentials, have exactly the same eigenvalues except one. Since we can find out the eigenvalues and eigenfunctions of the infinitely deep square well potential from any book on modern physics, SUSYQM automatically gives us all desired properties of its partner, the very nontrivial $(2\operatorname{cosec}^2 x - 1)$ potential. The eigenfunctions of the square well potential $V_-(x, 1)$, are given by $\psi_n^-(x, 1) \sim \sin(nx)$ and the corresponding eigenvalues by $E_n^- = n^2$, $n = 0, 1, 2, \ldots$. Hence, using $\psi_{n-1}^+ \sim A^-\psi_n^-$ and $A^- = \left(\frac{d}{dx} - \cot x\right)$, the eigenspectrum of the $\operatorname{cosec}^2 x$ potential is given by $\psi_{n-1}^+(x, 1) \sim \left(\frac{d}{dx} - \cot x\right)\sin(nx)$ and $E_n^+ = n^2$, $n = 1, 2, 3, \ldots$. Thus, this method gives a step-by-step simple method for recursively generating all eigenfunctions.

Problem 5.5. *Determine the following two functions and plot them:* $\psi_3^-(x, 1)$ *and* $\psi_2^+(x, 1)$.

Problem 5.6. *By explicitly operating with A^- and A^+ on eigenfunctions $\psi_3^-(x, 1)$ and $\psi_2^+(x, 1)$, verify that A^- removes a node while A^+ adds a node to the eigenfunctions of H_- and H_+.*

Problem 5.7. *The Dirac-delta function potential can be described by a superpotential $W(x) = \Theta(x - x_0) - \frac{1}{2}$.*

(1) Does this superpotential lead to unbroken or broken supersymmetry?

(2) *Determine the eigenfunction corresponding to the zero energy ground state.*

(3) *From the shape of the partner potential V_+, determine the number of bound states supported by an attractive Delta function potential.*

Chapter 6

Shape Invariance

In Chapter 5, we learned that a quantum mechanical system generated by a superpotential $W(x, a)$ yields two partner hamiltonians, $H_-(x, a)$ and $H_+(x, a)$. SUSYQM formalism relates these two hamiltonians in such a way that if we knew the spectrum of one, we could find the spectrum of the other.

6.1   Shape Invariant Potentials

There is a restricted class of superpotentials $W(x, a)$ for which the partner potentials $V_\pm(x, a)$ obey a special constraint known as "shape invariance." For this class we do not have to know the spectrum of one partner hamiltonian to know the other.

The shape invariance condition (also sometimes known as the property of reparametrizational invariance) states that a potential $V_+(x, a_0)$ can be written as $V_-(x, a_1)$ plus an additive constant $R(a_0)$. In other words, the partner potentials are related by

$$V_+(x, a_0) = V_-(x, a_1) + R(a_0) . \tag{6.1}$$

Parameters "a_0, a_1" represent characteristics of the potential, such as its strength, depth or width. We need to consider a family of potentials that differ from each other in the values of parameters a in them. Two shape invariant partner potentials $V_+(x, a_0)$ and $V_-(x, a_1)$ have the same x-dependence but differ from each other in the value of parameters a_i.

In terms of hamiltonians, the shape invariance condition reads

$$H_+(x, a_0) + g(a_0) = H_-(x, a_1) + g(a_1) , \tag{6.2}$$

where, we have written $R(a_0) = g(a_1) - g(a_0)$. Since these *two hamiltonians differ by a constant, their eigenvalues differ by the same constant and both have the same set of eigenfunctions.* I.e., $E_n^+(a_0) = E_n^-(a_1) + g(a_1) - g(a_0)$ and $\psi_n^+(x, a_0) = \psi_n^-(x, a_1)$ for all values of n.

Now, let us see how shape invariance leads to exact solvability. Suppose we want to find the spectrum of $H_-(x, a_0)$. Let us assume that our super-potentials $W(x, a_0)$ and $W(x, a_1)$ are such that $\psi_0^-(x, a_i) \sim e^{-\int_{x_0}^x W(x, a_i)\, dx}$ are normalizable functions, i.e., we have unbroken supersymmetry for the values of parameters a_i, $i = 0, 1, \ldots n$. If wave functions are not normalizable, supersymmetry will be broken. Solvability of systems with broken supersymmetry will be considered in Chapter 7.

Since supersymmetry is unbroken, we know that the ground state energy $E_0^-(a_i) = 0$, $i = 0, 1, \ldots, n$. From the shape invariance condition Eq. (6.2), we then know that $E_0^+(a_0) = E_0^-(a_1) + g(a_0) - g(a_1) = g(a_0) - g(a_1)$. From the isospectrality condition described in the previous chapter, we have $E_1^-(a_0) = E_0^+(a_0)$. Thus, we already have the two lowest eigenvalues of H_-: $E_0^-(a_0) = 0$ and $E_1^-(a_0) = g(a_0) - g(a_1)$.

Now let us compute the next eigenstate $E_2^-(a_0)$. From our discussion on isospectrality, we know that $E_2^-(a_0)$ is equal to $E_1^+(a_0)$. But from shape invariance condition we have $E_1^+(a_0) = E_1^-(a_1) + g(a_1) - g(a_0)$. Using isospectrality again, we have $E_1^-(a_1) = E_0^+(a_1)$, which in turn from shape invariance is equal to $E_0^-(a_2) + g(a_2) - g(a_1)$, where a_2 is another parameter that is related to a_1 in exactly the same way as a_1 is related to a_0; i.e., it is given by $f(a_1)$.

Thus we have

$$
\begin{aligned}
E_2^-(a_0) &\overset{\text{SUSYQM}}{=} E_1^+(a_0) \\
&\overset{\text{SI}}{=} E_1^-(a_1) + g(a_1) - g(a_0) \\
&\overset{\text{SUSYQM}}{=} E_0^-(a_2) + g(a_2) - g(a_1) + g(a_1) - g(a_0) \\
&= g(a_2) - g(a_0) ,
\end{aligned}
\tag{6.3}
$$

where we have assumed that the hamiltonian $H_-(x, a_2)$ also corresponds to unbroken SUSY and hence the ground state energy of the system $E_0^-(a_2) = 0$. The general expression for an eigenvalue of $H_-(x, a_0)$ is $E_n^-(a_0) = g(a_n) - g(a_0)$. Now let us find the eigenfunctions. The zero-energy ground state eigenfunction and the first excited state of H_- are given by

$$
\psi_0^-(x, a_0) \sim e^{-\int_{x_0}^x W(x, a_0)\, dx} , \quad \text{and}
$$
$$
\psi_1^-(x, a_0) \sim A^+(x, a_0)\, \psi_0^+(x, a_0).
\tag{6.4}
$$

However, from shape invariance we have $\psi_0^+(x, a_0) = \psi_0^-(x, a_1) \sim e^{-\int_{x_0}^x W(x, a_1)\, dx}$. Thus, we have

$$
\psi_1^-(x, a_0) \sim A^+(x, a_0) e^{-\int_{x_0}^x W(x, a_1)\, dx} .
\tag{6.5}
$$

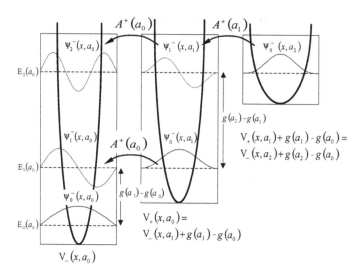

Fig. 6.1 Shape invariant potentials, their energy levels and the corresponding eigenfunctions.

Next, let us find $\psi_2^-(x, a_0)$. The ground state eigenfunction of $H_-(x, a_2)$ is given by $\psi_0^-(x, a_2) \sim \exp\left(-\int_{x_0}^x W(x, a_2)dx\right)$. Using Eq. (6.4), the corresponding eigenfunction is obtained as follows:

$$
\begin{aligned}
\psi_2^-(x, a_0) &= \frac{1}{\sqrt{E_1^+(a_0)}} A^+(x, a_0)\, \psi_1^+(x, a_0) \\
&= \frac{1}{\sqrt{E_1^+(a_0)}} A^+(x, a_0)\, \psi_1^-(x, a_1) \\
&= \frac{1}{\sqrt{E_1^+(a_0) E_0^+(a_1)}} A^+(x, a_0) A^+(x, a_1)\, \psi_0^+(x, a_1) \\
&= \frac{1}{\sqrt{E_1^+(a_0) E_0^+(a_1)}} A^+(x, a_0) A^+(x, a_1)\, \psi_0^-(x, a_2) \ . \quad (6.6)
\end{aligned}
$$

All eigenfunctions and eigenvalues of the hamiltonian $H_-(x, a_0)$ can be determined by this iterative algorithm. Thus, SUSYQM and shape invariance determine the entire spectrum of solvable quantum systems through a step-by-step recursive procedure, without any need to solve a differential equation.

6.2   Examples of Shape Invariant Systems

Earlier in Chapter 5 we used the spectrum of the infinite square well, which we knew from Modern Physics, to determine the spectrum of its partner, the $\operatorname{cosec}^2 x$ (Pöschl-Teller) potential. Now we demonstrate that they are in fact shape invariant partners and we can thus determine the spectrum of either without *a priori* knowledge of the other.

6.2.1   *Pöschl-Teller Potential Revisited*

The potential $V_+(x, b)$, given by

$$V_+(x,b) = W^2(x) + \frac{dW}{dx} = b(b+1)\operatorname{cosec}^2 x - b^2 \ , \qquad (6.7)$$

can be rewritten as

$$\begin{aligned} V_+(x,b) &= (b+1)[(b+1)-1]\operatorname{cosec}^2 x - (b+1)^2 + (b+1)^2 - b^2 \\ &= V_-(x,b+1) + (b+1)^2 - b^2 \ . \end{aligned} \qquad (6.8)$$

So the potential $V_-(x, b)$ is a shape invariant potential as defined in Eq. (6.1), with $a_0 = b$ and $a_1 = a_0 + 1 = b + 1$. Therefore, $g(b) = b^2$. Here, as a special case of the Pöschl-Teller potential, we solve the infinite square well. As we saw earlier, setting $b = 1$ leads to $V_+(x, 1) = 2\operatorname{cosec}^2 x - 1$ and $V_-(x, 1) = -1$, with the latter representing an infinitely deep potential well in the region $0 < x < \pi$. The ground state eigenfunction of $H_-(x, 1)$ is $\psi_0^-(x, 1) \sim e^{-\int W(x,1)dx} \sim e^{\int \cot x \, dx} \sim \sin x$; its energy is, as usual, 0. Now, we use shape invariance to determine the excited states.

From Eq. (6.8), $V_+(x, 1) = V_-(x, 2) + 3$; thus, $g(2) - g(1) = 3$. Using the fact that the ground state energy $E_0^-(2)$ of $H_-(x, 2)$ is zero (like all H_- ground states where ψ_0^- is normalizable), we find that the ground state energy of $H_+(x, 1)$ is $E_0^+(1) = 3$. But this is just the energy of the first excited state of H_-.

The common ground state eigenfunction of $H_+(x, 1)$ and $H_-(x, 2)$ is given by $\psi_0^+(x, 1) = \psi_0^-(x, 2) \sim e^{-\int W(x,2)dx} \sim e^{\int 2\cot x \, dx} \sim \sin^2 x$. Then the first excited state of $H_-(x, 1)$ is given by $\psi_0^-(x, 1) \sim A^+(x, 1)\sin x = \left(-\frac{d}{dx} - \cot x\right)\sin^2 x \sim \sin 2x$. Thus, we have derived the energy and the eigenfunction of the first excited state of $H_-(x, 1)$. By iterating this procedure, we can generate the entire spectrum. In particular, the eigenvalues are given by $E_n^-(b) = (b+n)^2 - b^2$.

In the above example, our parameters from Eqs. (6.1) and (6.2) were $a_0 = b$ and $a_1 = b + 1 = a_0 + 1$. This type of shape invariance, for which

$a_n = a_{n-1}$ + constant, is known as additive shape invariance. In contrast, shape invariance conditions in which parameters are related by $a_1 = q\,a_0$ are called multiplicative shape invariance. Potentials with multiplicative shape invariance are only known as series, not in closed form, and we will discuss them later.

6.2.2 *Harmonic Oscillator*

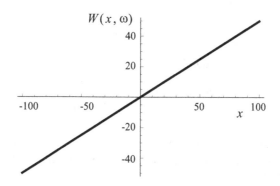

Fig. 6.2   Superpotential for one dimensional harmonic oscillator for $\omega = 1$.

This system is described by a superpotential $W(x,\omega) = \frac{1}{2}\omega x$ as shown in Figure 6.2. First of all we see that the superpotential is an odd function of x, hence we expect the ground state wave function to be normalizable. The wave function $\psi_0(x)$ is now given by $N\exp\left(-\int_0^x W(x,\omega)\,dx\right) = N\exp\left(-\int_0^x \frac{1}{2}\omega x\,dx\right) = N\exp\left(-\frac{1}{4}\omega x^2\right)$, where N is the normalization constant. Clearly, this function goes to zero at $\pm\infty$ and can be integrated over the entire domain to give a finite result. Now let us turn to its shape invariance. The partner potentials are now given by

$$V_\pm(x,\omega) = \frac{1}{4}\omega x^2 \pm \frac{1}{2}\omega\,.$$

Hence, we have

$$V_+(x,\omega) = V_-(x,\omega) + \omega\,.$$

In this case $g(a_1) - g(a_0) = \omega$, and $g(a_n) - g(a_0) = n\omega$. This is the origin of equal spacing of the eigenvalues of the harmonic oscillator.

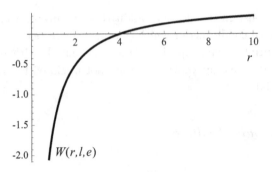

Fig. 6.3   Coulomb superpotential with $\ell = 1, e = 1$.

6.2.3   *Coulomb Potential*

The superpotential for the Coulomb system is

$$W(r, \ell, e) = \frac{1}{2} \frac{e^2}{\ell + 1} - \frac{\ell + 1}{r} \ .$$

It represents a particle with angular momentum ℓ interacting with a force center via an attractive Coulomb potential. An example for $\ell = 1$ is shown in Figure 6.3. The partner potentials are

$$V_+(r, \ell, e) = \frac{1}{4} \left(\frac{e^2}{\ell + 1} \right)^2 - \frac{e^2}{r} + \frac{(\ell + 1)(\ell + 2)}{r^2}$$

and   (6.9)

$$V_-(r, \ell, e) = \frac{1}{4} \left(\frac{e^2}{\ell + 1} \right)^2 - \frac{e^2}{r} + \frac{\ell(\ell + 1)}{r^2} \ .$$

This leads to

$$V_+(r, \ell, e) = V_-(r, \ell + 1, e) + \underbrace{\frac{1}{4} \left(\frac{e^2}{\ell + 1} \right)^2}_{-g(\ell + 1)} - \underbrace{\frac{1}{4} \left(\frac{e^2}{\ell + 2} \right)^2}_{-g(\ell + 2)} \ . \qquad (6.10)$$

Thus, with $a_0 = \ell + 1$ and $a_1 = \ell + 2$, the we have a shape invariant superpotential. $g(a_0)$ is $-\frac{1}{4} \left(\frac{e^2}{\ell + 1} \right)^2$ and the energy eigenvalues are given by

$$E_n^{(-)} = g(a_n) - g(a_0) = \frac{1}{4} \left(\frac{e^2}{\ell + 1} \right)^2 - \frac{1}{4} \left(\frac{e^2}{\ell + n + 1} \right)^2 \ .$$

The ground state eigenfunction is

$$\psi_0^{(-)}(r, \ell, e) = r^{\ell + 1} \exp \left(-\frac{1}{2} \frac{e^2}{\ell + 1} r \right) \ .$$

Problem 6.1. *Compute its first three excited state eigenfunctions, and determine for what values of its parameters is the supersymmetry unbroken?*

6.2.4 *The Hulthen Potential*

Another example of a shape invariant potential is the Hulthen potential:

$$V(r,a) = -\frac{Z}{a}\frac{\exp\left(-\frac{r}{a}\right)}{1-\exp\left(-\frac{r}{a}\right)} \ . \tag{6.11}$$

It is a short-range potential. For small values of r, i.e., $r \ll a$, the potential behaves as $-\frac{Z}{r}$; i.e., like an attractive Coulomb potential. Thus, the constant Z can be identified with the atomic number. For large values of r, the potential decreases exponentially: $V(r,a) = -\frac{Z}{a}\exp\left(-\frac{r}{a}\right)$. The Hulthen potential is used in many branches of physics, such as nuclear physics, atomic physics, solid state physics, and chemical physics. The model of the three-dimensional delta-function could well be considered as a Hulthen potential with the radius of the force going down to zero.

For the superpotential

$$W(r,l) = \frac{a}{l} - \frac{l}{a}\left[\frac{1+e^{-\frac{r}{a}}}{1-e^{-\frac{r}{a}}}\right] = \frac{a}{l} - \frac{l}{2a}\coth\left(r/2a\right) , \tag{6.12}$$

the corresponding partner potentials are

$$V_{\pm}(r,l) = \frac{l(l\pm 1)}{4a^2}\operatorname{cosech}^2\left(\frac{r}{2a}\right) - 2\coth\left(\frac{r}{2a}\right) + \left(\frac{l}{2a}\right)^2 + \left(\frac{a}{l}\right)^2 .$$

After some algebra, we find that

$$V_+(r,l) = V_-(r,l+1) + \underbrace{\left(\frac{l^2}{4a^2}+\frac{a^2}{l^2}\right)}_{-g(l)} - \underbrace{\left(\frac{(l+1)^2}{4a^2}+\frac{a^2}{(l+1)^2}\right)}_{-g(l+1)} .$$

Thus, we see that the superpotential given above leads to a shape invariant system. The potential $V_-(r,l)$, apart from an additive constant, reduces to the Hulthen potential for $l = 1$. Thus we obtained the entire spectrum of the potential without recourse to the differential equation.

Problem 6.2. *Obtain expressions for the three lowest eigenfunctions and the general expression for eigenvalues of the Hulthen potential given above.*

Problem 6.3. *For $W(x,a) = a\tanh x$,*

(1) show that it is a shape invariant potential;
(2) determine its eigenenergies and first two eigenfunctions;
(3) show that it holds a finite number of bound states.

Problem 6.4. *The following shape invariant superpotential*

$$W(x, a, b) = a \tan x - b \cot x ,$$

with $0 < x < \frac{\pi}{2}$ and $a > 0, b > 0$ is equivalent to

$$W(x, c, d) = c \operatorname{cosec} z + d \cot z .$$

Determine parameters c, d and the new variable z in terms of a, b and x. What is domain of the new superpotential?

We list below several translational shape invariant potentials, and their corresponding potentials and parameters.

- **Harmonic oscillator**

Superpotential:	$W = \frac{1}{2} \omega x , \quad -\infty < x < \infty$
Potential:	$V_- = \frac{1}{4} \omega x^2 - \frac{1}{2} \omega$
a_0:	a_0
a_1:	$a_0 + 1$
$g(a_0)$:	$g(a_0) = \omega a_0$
Eigenenergies:	$E_n^{(-)} = n\omega$

- **3-D Oscillator**

Superpotential:	$W = \frac{1}{2} \omega r - \frac{\ell+1}{r} , \quad 0 < r < \infty$
Potential:	$V_- = \frac{\ell(\ell+1)}{r^2} - \frac{(2\ell+3)\omega}{2} + \frac{\omega^2 r^2}{4}$
a_0:	a_0
a_1:	$a_0 + 1$
$g(a_0)$:	$g(a_0) = 2\omega a_0$
Eigenenergies:	$E_n^{(-)} = 2n\omega$

- **Coulomb**

Superpotential:	$W = \frac{e^2}{2(\ell+1)} - \frac{\ell+1}{r} , \quad 0 < r < \infty$
Potential:	$V_- = \frac{\ell(\ell+1)}{r^2} - \frac{e^2}{r} + \frac{e^4}{4(\ell+1)^2}$
a_0:	ℓ
a_1:	$\ell + 1$
$g(a_0)$:	$g(a_0) = -\frac{e^4}{4(a_0+1)^2}$
Eigenenergies:	$E_n^{(-)} = \frac{1}{4} \left(\frac{e^2}{\ell+1} \right)^2 - \frac{1}{4} \left(\frac{e^2}{\ell+n+1} \right)^2$

- **Morse**

Superpotential:	$W = A - e^{-x}$, $\quad -\infty < x < \infty$
Potential:	$V_- = A^2 + e^{-2x} - (2A+1)e^{-x}$
a_0:	A
a_1:	$A - 1$
$g(a_0)$:	$g(a_0) = -a_0^2$
Eigenenergies:	$E_n^{(-)} = A^2 - (A-n)^2$

- **Scarf (hyperbolic)**

Superpotential:	$W = A \tanh x + B \operatorname{sech} x$, $\quad -\infty < x < \infty$
Potential:	$V_- = -\left[A(A+1) - B^2\right] \operatorname{sech}^2 x$
	$\quad\quad + 2(A+1)B \tanh x \operatorname{sech} x + A^2$
a_0:	A
a_1:	$A - 1$
$g(a_0)$:	$g(a_0) = -a_0^2$
Eigenenergies:	$E_n^{(-)} = A^2 - (A-n)^2$

- **Rosen-Morse (hyperbolic)**

Superpotential:	$W = A \tanh x + \frac{B}{A}$, $\quad -\infty < x < \infty$, $B < A^2$
Potential:	$V_- = -A(A+1) \operatorname{sech}^2 x + 2B \tanh x + A^2 + \frac{B^2}{A^2}$
a_0:	A
a_1:	$A - 1$
$g(a_0)$:	$g(a_0) = -a_0^2 - \frac{B^2}{a_0^2}$
Eigenenergies:	$E_n^{(-)} = A^2 - (A-n)^2 - \frac{B^2}{(A-n)^2} + \frac{B^2}{A^2}$

- **Eckart**

Superpotential:	$W = -A \coth r + \frac{B}{A}$, $\quad 0 < r < \infty$, $B > A^2$
Potential:	$V_- = A(A-1) \operatorname{cosech}^2 r - 2B \coth r + A^2 + \frac{B^2}{A^2}$
a_0:	A
a_1:	$A + 1$
$g(a_0)$:	$g(a_0) = -a_0^2 - \frac{B^2}{a_0^2}$
Eigenenergies:	$E_n^{(-)} = A^2 - (A+n)^2 - \frac{B^2}{(A+n)^2} + \frac{B^2}{A^2}$

- **Pösch-Teller**

  Superpotential: $W = A \coth r - B \operatorname{cosech} r$, $0 < r < \infty$, $A < B$

  Potential: $V_- = \left[A(A+1) + B^2 \right] \operatorname{cosech}^2 r + A^2$
  $$-(2A+1)B \coth r \operatorname{cosech} r$$

  a_0: A

  a_1: $A - 1$

  $g(a_0)$: $g(a_0) = -a_0^2$

  Eigenenergies: $E_n^{(-)} = A^2 - (A-n)^2$

- **Scarf (trigonometric)**

  Superpotential: $W = A \tan x - B \sec x$, $-\frac{\pi}{2} < x < \frac{\pi}{2}$, $A > B$

  Potential: $V_- = \left[A(A-1) + B^2 \right] \sec^2 x - A^2$
  $$-(2A-1)B \tan x \sec x$$

  a_0: A

  a_1: $A + 1$

  $g(a_0)$: $g(a_0) = a_0^2$

  Eigenenergies: $E_n^{(-)} = (A+n)^2 - A^2$

- **Rosen-Morse (trigonometric)**

  Superpotential: $W = -A \cot x - \frac{B}{A}$, $0 < x < \frac{\pi}{2}$

  Potential: $V_- = A(A-1) \operatorname{cosec}^2 x + 2B \cot x - A^2 + \frac{B^2}{A^2}$

  a_0: A

  a_1: $A + 1$

  $g(a_0)$: $g(a_0) = -\frac{B^2}{a_0^2} + a_0^2$

  Eigenenergies: $E_n^{(-)} = (A+n)^2 - A^2 + \frac{B^2}{A^2} - \frac{B^2}{(A+n)^2}$

Chapter 7

Supersymmetry and Its Breaking

In Chapters 5 and 6, we briefly discussed cases of broken and unbroken supersymmetry. In this chapter we will describe this concept in detail and discuss its repercussions.

We have seen that in supersymmetric quantum mechanics the partner hamiltonians are given by $H_- = A^+ A^-$ and $H_+ = A^- A^+$, where A^+ and A^- are hermitian conjugate operators of each other. The structure of these hamiltonians guarantees that their eigenenergies are never negative.

$$\langle \phi | H_\mp | \phi \rangle \equiv \langle \phi | A^\pm A^\mp | \phi \rangle = \left(\langle \phi | A^\pm \right) \left(A^\mp | \phi \rangle \right)$$
$$= \left| A^\mp | \phi \rangle \right|^2 \geq 0 \ . \tag{7.1}$$

Since the above statement is true for an arbitrary state of the system, for the ground state we know that the energy is either zero or positive. We will soon see the significance of these two possibilities.

Let us explore the conditions under which we have unbroken or broken supersymmetry. For that let us introduce two new operators $Q^\pm$ that will allow us to express many of the equations we saw earlier in a matrix form:

$$Q^- \equiv \begin{pmatrix} 0 & 0 \\ A^- & 0 \end{pmatrix} \quad \text{and} \quad Q^+ \equiv \begin{pmatrix} 0 & A^+ \\ 0 & 0 \end{pmatrix} \ . \tag{7.2}$$

Multiplying these operators we obtain

$$Q^- Q^+ = \begin{pmatrix} 0 & 0 \\ 0 & A^- A^+ \end{pmatrix} \quad \text{and} \quad Q^+ Q^- = \begin{pmatrix} A^+ A^- & 0 \\ 0 & 0 \end{pmatrix} \ . \tag{7.3}$$

We can now define a hamiltonian H in matrix form with $H_\mp$ as the diagonal components:

$$H \equiv \begin{pmatrix} H_- & 0 \\ 0 & H_+ \end{pmatrix} = \begin{pmatrix} A^+ A^- & 0 \\ 0 & A^- A^+ \end{pmatrix} \ . \tag{7.4}$$

However, the above hamiltonian is just: $Q^-Q^+ + Q^+Q^-$. This expression is known as the anti-commutator of operators $Q^\pm$ and is denoted by $\{Q^-, Q^+\}$. Thus, we have

$$\{Q^-, Q^+\} \equiv Q^-Q^+ + Q^+Q^- = H \ . \tag{7.5}$$

Problem 7.1.

(1) Show that $(Q^\pm)^2 = 0$.
(2) Show that operators $Q^\pm$ commute with hamiltonian H; i.e.

$$[Q^\pm, H] \equiv Q^\pm H - H Q^\pm = 0 \ .$$

The first property

$$[Q^\pm, H] \ = \ 0 \ , \tag{7.6}$$

signals the presence of an underlying symmetry in the system. Explicitly it implies that $Q^- H |\psi\rangle - H Q^- |\psi\rangle = 0$, where $|\psi\rangle$ is an eigenstate of the hamiltonian. Thus $H (Q^- |\psi\rangle) = E (Q^- |\psi\rangle)$; i.e., the operation of $Q^\pm$ on state $|\psi\rangle$ does not change the energy of the state, and hence represents a symmetry of the system. This is the supersymmetry of SUSYQM, and $Q^\pm$ are called its generators.

The second property of operators

$$(Q^-)^2 = (Q^+)^2 = 0 \ , \tag{7.7}$$

implies that the action of operators $Q^\pm$ twice on any state does not generate a new state, as its magnitude is zero. This reminds one of the Pauli exclusion principle where any attempt to create two fermions in a state is bound to be futile. Because of this behavior of $Q^\pm$ we term them fermionic.

In Chapter 5 we learned that supersymmetry operators $A^\pm$ interconnect eigenstates of $H_\pm$. Let us now see how these interconnections are represented in terms of operators $Q^\pm$ and H. Eigenfunctions $\psi_n^-(x)$ and $\psi_{n-1}^+(x)$ of hamiltonians $H_\mp$ can be recast into column-vector form, $\begin{pmatrix} \psi_n^-(x) \\ 0 \end{pmatrix}$ and $\begin{pmatrix} 0 \\ \psi_{n-1}^+(x) \end{pmatrix}$, and they are then eigenstates of the hamiltonian H.

Problem 7.2. *Verify that vectors $\begin{pmatrix} \psi_n^-(x) \\ 0 \end{pmatrix}$ and $\begin{pmatrix} 0 \\ \psi_{n-1}^+(x) \end{pmatrix}$ are degenerate eigenstates of the hamiltonian H with eigenvalue $E_n^- = E_{n-1}^+$.*

These column vectors are related through the charges $Q^\mp$ as follows:

$$Q^- \begin{pmatrix} \psi_n^-(x) \\ 0 \end{pmatrix} \equiv \begin{pmatrix} 0 & 0 \\ A^- & 0 \end{pmatrix} \begin{pmatrix} \psi_n^-(x) \\ 0 \end{pmatrix}$$

$$= \begin{pmatrix} 0 \\ A^- \psi_n^-(\dot{x}) \end{pmatrix} = \begin{pmatrix} 0 \\ \psi_{n-1}^+(x) \end{pmatrix}, \qquad (7.8)$$

and

$$Q^+ \begin{pmatrix} 0 \\ \psi_n^+(x) \end{pmatrix} \equiv \begin{pmatrix} 0 & A^+ \\ 0 & 0 \end{pmatrix} \begin{pmatrix} 0 \\ \psi_n^+(x) \end{pmatrix}$$

$$= \begin{pmatrix} A^+ \psi_n^+(x) \\ 0 \end{pmatrix} = \begin{pmatrix} \psi_{n+1}^-(x) \\ 0 \end{pmatrix}. \qquad (7.9)$$

Since $Q^\pm$ do not change the energy of a state, these transformations relate two states with the same eigenvalue of H. Thus, while $A^\pm$ connected eigenstates of two different hamiltonians, operators $Q^\pm$ transform two states of H into each other and hence demonstrate the degeneracy of the system.

We know from spin-orbit coupling that integer orbital angular momentum ℓ when added to spin angular momentum $s = \frac{1}{2}$, gives us total angular momentum $j = \ell \pm \frac{1}{2}$. Recalling that integer momentum states are associated with bosons and half-integer states with fermions, we recognize that a coupling between a boson and a fermion always leads to a fermion. Analogously, we expect that operators $Q^\pm$, which we established to have fermionic behavior, applied to a state would change its character from bosonic to fermionic and vice-versa. However, here the concept of bosonic or fermionic is rather abstract. We consider, arbitrarily, the states of the form $\begin{pmatrix} \psi_n^-(x) \\ 0 \end{pmatrix}$ to be bosonic. Then $Q^- \begin{pmatrix} \psi_n^-(x) \\ 0 \end{pmatrix} = \begin{pmatrix} 0 \\ \psi_{n-1}^+(x) \end{pmatrix}$ are fermionic states. Similarly, when Q^+ is applied to a fermionic state $\begin{pmatrix} 0 \\ \psi_n^+(x) \end{pmatrix}$, it gives a state of the form $\begin{pmatrix} \psi_{n+1}^-(x) \\ 0 \end{pmatrix}$, a bosonic state. In short, vectors of the form $\begin{pmatrix} \phi(x) \\ 0 \end{pmatrix}$ and $\begin{pmatrix} 0 \\ \phi(x) \end{pmatrix}$ represent a bosonic and a fermionic state respectively.

In summary, we have defined supersymmetry generators $Q^\pm$. Their anticommutator generates the hamiltonian of the system. Since these operators commute with H, there is supersymmetry in the system.

7.1   Breaking of Supersymmetry

Despite this symmetry structure at the level of the operators, it is possible to have a system where the symmetry is broken. Let us consider the case that supersymmetry is the symmetry of the hamiltonian; i.e., $HQ^\pm - Q^\pm H = 0$. If this were not so we would say that supersymmetry is explicitly broken. But even if H commutes with $Q^\pm$, there is another kind of symmetry breaking known as the spontaneous breaking of supersymmetry. Let us see how that works.

A unitary operator [1] that generates the supersymmetry transformation of a state is of the form $e^{i\left(\epsilon Q^- + \bar{\epsilon} Q^+\right)}$, where ϵ is a parameter that measures how far the state deviates from the original state [2] and $\bar{\epsilon}$ is its complex conjugate. For $\epsilon = 0$, we obtain the identity, as expected.

The ground state of the system (also sometimes known as the vacuum state), represented by $|0\rangle$, must be unique. It must not change when operated upon by $e^{i\left(\epsilon Q^- + \bar{\epsilon} Q^+\right)}$. In other words, the ground state must be invariant under this symmetry operation. I.e., if supersymmetry is to remain a good symmetry of the system, for an arbitrary value of the parameter ϵ, we must have

$$e^{i\left(\epsilon Q^- + \bar{\epsilon} Q^+\right)} |0\rangle = |0\rangle \ . \tag{7.10}$$

This implies that for unbroken SUSY we must have $Q^\pm |0\rangle = 0$. If $Q^\pm |0\rangle \neq 0$, i.e., if the ground state does not respect the symmetry, we have a case of spontaneous breakdown of supersymmetry.

Problem 7.3. *Show that* $\langle 0|H|0\rangle \neq 0$ *implies that either* $|Q^-|0\rangle| \neq 0$ *or* $|Q^+|0\rangle| \neq 0$, *and hence supersymmetry is broken.*

Since the hamiltonian H is given by the sum $(Q^- Q^+ + Q^+ Q^-)$, the supersymmetry is unbroken if and only if the energy of the ground state of hamiltonian H is zero. Non-vanishing of the ground state energy signals the breaking of SUSY.

If the supersymmetry is unbroken, we have

$$\begin{pmatrix} A^+ A^- & 0 \\ 0 & A^- A^+ \end{pmatrix} \begin{pmatrix} \phi^-(x) \\ \phi^+(x) \end{pmatrix} = 0. \tag{7.11}$$

---

[1] Unitary operators play a very important role in quantum mechanics. Their operation on a vector does not change the length of that vector. A unitary operator U obeys $U^\dagger U = U U^\dagger = 1$. Thus, the action by a unitary operator can be viewed as a generalization of multiplication by a complex number of modulus unity.

[2] Here the exponentiation of an operator is to be understood in terms of a power series. I.e. $e^A = 1 + A + \frac{1}{2!} A^2 + \frac{1}{3!} A^3 + \cdots$.

I.e., we must have $A^+ A^- \phi^-(x) = 0$, and $A^- A^+ \phi^+(x) = 0$. These in turn imply

$$A^- \phi^-(x) = 0, \quad \text{and} \quad A^+ \phi^+(x) = 0. \tag{7.12}$$

Problem 7.4. *Show that Eq. (7.11) leads to Eqs. (7.12).*

The solutions of Eqs. (7.12) are

$$\phi^{\pm}(x) = N_{\pm} \exp\left(\pm \int_{x_0}^{x} W(x', a_0)\, dx'\right) ;$$

where $N_{\pm}$ are constants of normalization. This implies $\phi^+(x) \sim 1/\phi^-(x)$. Hence, if $\phi^-(x)$ is normalizable; i.e., it goes to zero at infinity, the function $\phi^+(x)$ would not be a normalizable function as it would blow up at infinity. The only way to have a normalizable zero energy eigenstate is for either $\phi^-(x)$ or $\phi^+(x)$ to vanishe identically. Let us assume that it is the function $\phi^+(x)$ that is identically zero. Then the ground state of the system has the form:

$$\begin{pmatrix} \phi^-(x) \\ 0 \end{pmatrix} .$$

Since the function $\phi^-(x)$ is an eigenstate of the operator $A^+ A^-$ with eigenvalue zero, we will denote it by $\psi_0^{(-)}(x)$. This function is then given by

$$\psi_0^{(-)}(x) = N \exp\left(-\int_{x_0}^{x} W(x', a_0)\, dx'\right) .$$

As we saw in Chapter 5 normalizability of the state requires that wave function $\psi_0^{(-)}(x)$ vanishes at $x = \pm\infty$. This implies that $\int_{x_0}^{\pm\infty} W(x', a_0)\, dx'$ must be ∞. For the integral $\int_{x_0}^{\infty} W(x', a_0)\, dx'$ to be unbounded requires that function $W(x', a_0)$ must have a positive value as $x \to \infty$. Similarly, for $\int_{x_0}^{-\infty} W(x', a_0)\, dx'$ to be unbounded requires that function $W(x', a_0)$ must have a negative value as $x \to -\infty$. Thus, the function $W(x', a_0)$ must have an odd number of zeroes on the real axis for the normalizability of the ground state function $\psi_0^{(-)}(x)$, guaranteeing the existence of a zero energy state for H and the presence of supersymmetry. An odd function automatically guarantees all of above. In Figure 7.1, we show two examples of W's, one for infinite domain and other for a finite domain, that correspond to a system with zero energy ground state and hence unbroken SUSY.

On the other hand, if $W(x', a_0)$ has an even number of zeroes, or no zeroes at all, then SUSY is automatically broken spontaneously. Here in Figure 7.2 we depict some examples of $W(x', a_0)$ that lead to a system with spontaneously broken supersymmetry.

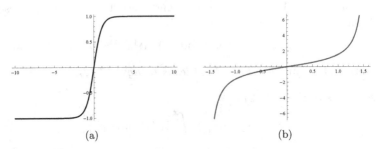

(a)   (b)

Fig. 7.1   Examples of superpotential $W(x', a_0)$ related to unbroken SUSY.

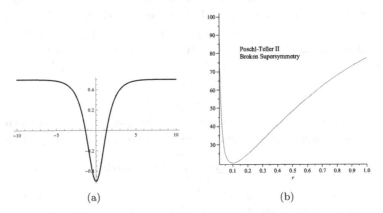

(a)   (b)

Fig. 7.2   Examples of superpotential $W(x', a_0)$ related to broken SUSY.

7.2   The Witten Index

In broken supersymmetric cases, there is no zero energy ground state. Hence the number of fermionic states N_F (eigenstates of H with upper component zero) is equal in number to those of the bosonic type N_B (eigenstates of H with lower component zero). In the unbroken case the number of bosonic states is one greater than that of fermionic states. Let us define an operator F that gives zero acting on a bosonic state and returns a fermionic state unchanged. Thus, bosonic and fermionic states would be eigenstates of operator $(-)^F$ with eigenvalues 1 and -1 respectively. Edward Witten [3] devised an index [4] $\Delta(\beta) = Tr(-)^F e^{-\beta H}$ that is zero if and only

---

[3] E. Witten, "Constraints on Supersymmetry Breaking", *Nucl. Phys.* **B 202**, 253 (1982).

[4] Trace Tr is the sum of all eigenvalues of an operator.

if the difference $N_B - N_F$ is zero. This index determines whether SUSY is unbroken or broken. For the broken case it returns 0; for the unbroken case it gives 1. Nontrivial application of this index are found in field theory.

Problem 7.5. *Show that* $N_B - N_F = 0$ *indeed implies that the index* $\Delta(\beta) = Tr\,(-)^F\,e^{-\beta H} = 0.$

Problem 7.6. *Show that* $N_B - N_F = 1$ *indeed implies that the index* $\Delta(\beta) = Tr\,(-)^F\,e^{-\beta H} = 1.$

7.3 Two-Step Shape Invariance for Systems with Broken Supersymmetry

In Chapter 6, the shape invariance of potentials was utilized to determine eigenvalues and eigenfunctions of various hamiltonians. This depend crucially on the system having a zero energy ground state; we could not have done this without supersymmetry. In this chapter we analyze shape invariant systems with broken supersymmetry. I.e., they are described by shape invariant potentials where values of the parameters leave the supersymmetry spontaneously broken. For these systems, the usual step-by-step procedure of Chapter 6 does not work.

At the outset, let us be clear that we are only interested in studying those potentials that admit bound states. In the list of potentials given in Chapter 6, only three of them hold bound states in the broken supersymmetric phase. For example, the Morse potential, described by the superpotential $W(x,\,A,\,B) = A - B\,e^{-x}$, does not hold bound states in the broken SUSY phase given by $A < 0$. This can be verified by drawing partner potentials for this superpotential for various values of A. It is important to note that since $A - B\,e^{-x} \equiv A - e^{-(x-x_0)}$, where $x_0 = \ln B$, the parameter B only moves the origin for x. It cannot affect the nature of supersymmetry.

We will now work with three potentials that hold bound states in their broken SUSY phase. These are: the three dimensional harmonic oscillator, the Pöschl-Teller I, and the Pöschl-Teller II.

There exists a variant of shape invariance that is referred to as a two-step shape invariance that will help us determine the spectrum and eigenstates for these systems with broken SUSY. The first step converts the initial superpotential with broken SUSY into one where SUSY is not broken. The second step uses the method of Chapter 6 to determine the spectrum. We will demonstrate this by examples.

7.3.1   *Two Step Shape Invariance via Examples*

We consider first the example of the Pöschl-Teller I superpotential

$$W(x, A, B) = A \tan x - B \cot x \; ; \quad 0 < x < \pi/2 \; . \qquad (7.13)$$

The supersymmetric partner potentials are given by

$$
\begin{aligned}
V_-(x, A, B) &= A(A-1) \; \sec^2 x \\
&\quad + B(B-1) \; \mathrm{cosec}^2 x - (A+B)^2 \; ; \qquad (7.14) \\
V_+(x, A, B) &= A(A+1) \; \sec^2 x + B(B+1) \; \mathrm{cosec}^2 x - (A+B)^2 \; .
\end{aligned}
$$

The zero energy ground state of the system, if normalizable, would be given by $\exp\left(-\int^x W(y, A, B) \; dy\right) \equiv \cos^A x \sin^B x$. For $A > 0, B > 0$, the function $\psi_0^{(-)}(x, A, B)$ is normalizable and proportional to the ground state wave function, implying unbroken SUSY. Similarly, for $A < 0, B < 0$, the function $1/\psi_0^{(-)}(x, A, B)$ is normalizable, and thus has unbroken SUSY.

For $A > 0$, and $B < 0$, $\psi_0^{(-)}(x, A, B)$ is not normalizable due to its divergent behavior at $x = 0$ while its reciprocal $\frac{1}{\psi_0^{(-)}}$ is not normalizable due to its divergent behavior at $x = \frac{\pi}{2}$, and hence SUSY is necessarily broken. Likewise, for $A < 0$, and $B > 0$, once again the SUSY is broken. As mentioned before, the broken SUSY case has no zero energy state, and the standard shape invariance procedure of Chapter 6 cannot be used to obtain the eigenstates and eigenvalues.

Let us focus on the case $A > 0$, $B < 0$. The eigenstates of $V_\pm(x, A, B)$, even without the presence of SI, are related by

$$
\begin{aligned}
\psi_n^{(+)}(x, a_0) &= A(x, a_0) \, \psi_n^{(-)}(x, a_0) \; ; \\
\psi_n^{(-)}(x, a_0) &= A^+(x, a_0) \, \psi_n^{(+)}(x, a_0)
\end{aligned}
\qquad (7.15)
$$

and

$$E_n^{(-)}(a_0) = E_n^{(+)}(a_0) \; .$$

The potentials of Eq. (7.14) are shape invariant. In fact there are two possible relations between parameters such that these two potentials exhibit shape invariance. One of them is the conventional $(A \rightarrow A+1, \; B \rightarrow B+1)$. The shape invariance condition is given by

$$V_+(x, A, B) = V_-(x, A+1, B+1) + (A+B+2)^2 - (A+B)^2 \; . \quad (7.16)$$

For B sufficiently large and negative, $B+1$ is also negative; thus the super-potential resulting from this change of parameters still falls in the broken SUSY category, and hence $E_0^{(-)}(a_0) \neq 0$. In the absence of this crucial result, even with shape invariance, we are not able to proceed further.

The second possibility is $(A \to A + 1, \quad B \to -B)$. The corresponding relationship is given by

$$V_+(x, A, B) = V_-(x, A + 1, -B) + (A - B + 1)^2 - (A + B)^2. \quad (7.17)$$

This change of parameters $(A \to A + 1, \quad B \to -B)$ leads to a system with unbroken SUSY, since the parameter B changes sign. Hence the ground state of the system with potential $V_-(x, A + 1, -B)$ is guaranteed to be at zero energy. From Eq. (7.17) we see that potentials $V_+(x, A, B)$ and $V_-(x, A + 1, -B)$ differ only by a constant, hence we have

$$\psi_+(x, A, B) = \psi_-(x, A + 1, -B) \ , \qquad \text{and}$$
$$E_n^{(+)}(A, B) = E_n^{(-)}(A + 1, -B) + (A + 1 - B)^2 - (A + B)^2. \quad (7.18)$$

Thus, if we knew the spectrum of the unbroken SUSY $H_-(x, A + 1, -B)$, we would be able to determine the spectrum of $H_+(x, A, B)$ with broken SUSY. Since the parameters of the potential $V_-(x, A + 1, -B)$ lie in the region necessary for unbroken SUSY, we have machinery at hand to determine the eigenstates of this potential. The results are

$$E_n^{(-)}(A + 1, -B) = (A + 1 - B + 2n)^2 - (A + 1 - B)^2. \quad (7.19)$$

When combined with Eqs. (7.15) and (7.17), we obtain

$$E_n^{(-)}(A, B) = (A + 1 - B + 2n)^2 - (A + B)^2 \ ; \qquad \text{and}$$
$$\psi_n^{(-)}(y, A, B) = (1 + y)^{(1-A)/2}(1 - y)^{B/2} P_n^{(B-1/2, \, 1/2 - A)}(y), \quad (7.20)$$

where $y = \cos(2x)$ and $P_n^{(\alpha, \, \beta)}(y)$ are Jacobi polynomials.

Similar analyses can be used for determining the eigenvalues and eigenstates of the three dimensional harmonic oscillator, and Pöschl-Teller II potentials.

7.3.1.1 *Three-Dimensional Harmonic Oscillator with Broken SUSY*

The three-dimensional harmonic oscillator with broken SUSY is described by the superpotential

$$W(r, \ell, \omega) = \frac{1}{2} \, \omega r - \frac{\ell + 1}{r} \ ; \quad \ell < -1 \ . \quad (7.21)$$

The function $\psi_0^{(-)}(r, \ell, \omega) = \exp\left(-\int^r W(r', \ell, \omega)\,dr'\right) = r^{\ell+1}e^{-\omega r^2}$ diverges near $r \to 0$ for $\ell < -1$ and hence corresponds to the case of broken SUSY. The supersymmetric partner potentials are

$$V_+(r, \ell, \omega) = \frac{\omega^2 r^2}{4} + \frac{(\ell+1)(\ell+2)}{r^2} - \left(\ell + \frac{1}{2}\right)\omega\ , \quad \text{and}$$

$$V_-(r, \ell, \omega) = \frac{\omega^2 r^2}{4} + \frac{\ell(\ell+1)}{r^2} - \left(\ell + \frac{3}{2}\right)\omega. \tag{7.22}$$

These two partner potentials are shape invariant since

$$V_+(r, \ell, \omega) = V_-(r, \ell+1, \omega) + 2\omega\ . \tag{7.23}$$

For sufficiently large negative values of ℓ, the potential $V_-(r, \ell+1, \omega)$ also lies in the realm of broken SUSY. However, there is another change of parameters that maintains shape invariance

$$V_+(r, \ell, \omega) = V_-(r, -\ell-2, \omega) - (2\ell+1)\omega\ . \tag{7.24}$$

Since $\ell < -1$, the potential $V_-(r, -\ell-2, \omega)$ has unbroken SUSY and thus has a zero energy ground state. Combining the results of Eqs. (7.22), (7.23) and (7.24), we obtain

$$E_n^{(+)}(\ell, \omega) = (2n - 2\ell - 1)\omega\ . \tag{7.25}$$

7.3.1.2 *Pöschl-Teller II Potential with Broken SUSY*

Our last example is the Pöschl-Teller II potential described by

$$W(r, A, B) = A\tanh r - B\coth r\ ; \quad 0 < r < \infty\ . \tag{7.26}$$

The function $\psi_0^{(-)}(r, A, B) \equiv \exp\left(-\int^r W(r', A, B)\,dr'\right)$ is given by $\psi_0^{(-)}(r, A, B) = \cosh^{-A} r \sinh^B r$. Here, for $A > 0$ and $B < 0$, neither $\psi_0^{(-)}(r, A, B)$ nor its inverse are normalizable, and hence we have a system with broken SUSY. The supersymmetric partner potentials are given by

$$V_+(r, A, B) = -A(A-1)\mathrm{sech}^2 r + B(B-1)\mathrm{cosech}^2 r + (A+B)^2\ ,$$
$$\text{and} \tag{7.27}$$
$$V_-(r, A, B) = -A(A+1)\mathrm{sech}^2 r + B(B+1)\mathrm{cosech}^2 r + (A+B)^2\ .$$

Here too we have two possible relations between parameters for these potentials to be shape invariant. They are

$$V_+(r, A, B) = V_-(r, A-1, B-1) + (A+B)^2 - (A+B-2)^2\ , \tag{7.28}$$

and

$$V_+(r,\, A,\, B) = V_-(r,\, A-1,\, -B) + (A+B)^2 - (A-B-1)^2 \,. \tag{7.29}$$

As in the previous two examples, the first transformation does not lead to the determination of the spectrum since the new parameters $(A-1, B-1)$ still leave the system with broken SUSY for sufficiently large positive value of A. However, in the second transformation, the new values of the parameters $(A-1,\ -B)$ lie in the domain of unbroken SUSY and hence the spectrum of $V_-(r, A-1, -B)$ can be determined. The resulting spectrum is given by

$$E_n^{(-)}(A,\, B) = (A+B)^2 - (A-1-B-2n)^2.$$

A similar procedure yields the eigenstates of the Pöschl-Teller II and harmonic oscillator potentials in the broken SUSY phase.

7.4 Breaking of Supersymmetry by Singular Potentials

In this section we will consider superpotentials that become singular over a region, depending on a parameter that ranges in value from finite to infinite. Our aim is to analyze how the supersymmetry in a system of this type is affected when the superpotential in a region is raised in magnitude so as to shut the wave function out of that region. The particular example that we will use for this study is the half-harmonic oscillator (HHO) defined over the region $[0, \infty)$ with an interaction of the type $\frac{1}{4}\omega x^2$. We will initially define a system over the entire real axis $(-\infty, \infty)$. Then by raising the value of a parameter λ, we will squeeze the wave function into the region $x \geq 0$, and thus reduce the system to that of a HHO.

In problems like this it is very important to understand what types of boundary conditions should be imposed on wave functions when the potential has a discontinuity in space. As we proceed, we will encounter many dilemmas and will need to make careful choices. Our approach to this problem will be guided by the wisdom gained from well known examples of quantum physics. Thus, before delving into the HHO, we will consider a problem that is very familiar to all of us; viz., the system of a particle in an infinitely deep well. We know its eigenvalues and eigenfunctions. We will construct an infinitely deep potential well starting from a finite well of depth Λ. As we take the $\Lambda \to \infty$ limit, we will ensure that our choices at every step lead to boundary conditions that do not exclude the well solutions of this system.

7.4.1   *Boundary Conditions for an Infinitely Deep Well*

We first start out with a finite well defined by

$$V(x) = \begin{cases} \Lambda \text{ for } x < 0 & \text{Region I} \\ 0 \text{ for } 0 < x < a & \text{Region II} \\ \Lambda \text{ for } x > a & \text{Region III} \end{cases} \quad . \qquad (7.30)$$

Wave functions in these three regions are given by

$$\begin{aligned} \psi_I(x) &= A e^{\kappa x} \\ \psi_{II}(x) &= B \sin k x + C \cos k x \\ \psi_{III}(x) &= D e^{-\kappa (x-a)} \end{aligned} \qquad (7.31)$$

respectively. Here

$$k = \sqrt{E} \quad \text{and} \quad \kappa = \sqrt{\Lambda - E} \; .$$

To determine the eigenspectrum of this system, we impose the following set of boundary conditions at $x = 0$.

$$\psi_I(0) = \psi_{II}(0) \quad \text{and} \quad \frac{d\psi_I(x)}{dx}\bigg|_{0-} = \frac{d\psi_{II}(x)}{dx}\bigg|_{0+} . \qquad (7.32)$$

These boundary conditions result in

$$A = C \quad \text{and} \quad \kappa A = k B . \qquad (7.33)$$

We impose similar boundary conditions at $x = a$. These relationship are only valid as long as Λ is finite. Let us analyze carefully what happens if we try to take the limit $\Lambda \to \infty$ to generate an infinitely deep potential well.

The Schrödinger equation is given by

$$-\frac{d^2\psi}{dx^2} + (V(x) - E)\,\psi = 0 . \qquad (7.34)$$

Consider a very small interval $[-\Delta, \Delta]$. Integrating the above equation equation from $x = -\Delta$ to Δ, we obtain

$$-\frac{d\psi}{dx}\bigg|_{-\Delta}^{+\Delta} + \int_{-\Delta}^{+\Delta} (V(x) - E)\,\psi\,dx = 0 \; . \qquad (7.35)$$

For infinitesimal values of Δ we can ignore the integral $\int_{-\Delta}^{+\Delta} E\psi\,dx$ since E is constant and ψ is well-behaved. Then Eq. (7.35) reduces to

$$-\frac{d\psi}{dx}\bigg|_{-\Delta}^{+\Delta} + \int_{-\Delta}^{+\Delta} V(x)\,\psi\,dx = 0 \; . \qquad (7.36)$$

Now substituting the explicit form of the wave function, near the $x = 0$ boundary, we obtain

$$-\left.\frac{d\psi}{dx}\right|_{-\Delta}^{+\Delta} + A\ \Lambda \int_{-\Delta}^{0} e^{\kappa x}\ dx = 0\ .$$

Evaluating the integral, we obtain

$$\left.\frac{d\psi}{dx}\right|_{-\Delta}^{+\Delta} = \frac{A\ \Lambda}{\kappa}\left(1 - e^{-\kappa\Delta}\right)\ ,$$

which for $\Lambda \gg E$, yields

$$\left.\frac{d\psi}{dx}\right|_{-\Delta}^{+\Delta} = A\ \kappa\ \left(1 - e^{-\kappa\Delta}\right)\ . \tag{7.37}$$

In order to obtain the correct solutions for an infinitely deep potential well, there are two independent limiting procedures which will have to handled carefully: $\kappa \to \infty$ and $\Delta \to 0$. In Eq. (7.37), if we take the limit $\Delta \to 0$ first, we obtain eigenfunctions of "an infinitely deep potential well" with zero slopes at boundaries, which is wrong. On the other hand, if we take $\kappa \to \infty$-limit first, we see that Eq. (7.37) can generate the familiar finite discontinuity in the derivative, provided $A \sim \frac{1}{\kappa}$. (The magnitude of the discontinuity can be determined by the additional condition of normalization.) For infinitely large values of κ, this then leads to $A \to 0$, $C \to 0$ and finite B, thus producing correct solutions for an infinitely deep potential well.

Thus, taking $\Delta \to 0$ first leads to incorrect eigenfunctions for the infinite well, while taking $\kappa \to 0$ first provides eigenfunctions for the infinite potential well together with the necessary discontinuity in the derivative, and vanishing of the wave function at the boundary.

7.4.2 *Boundary Conditions for a Half-Harmonic Oscillator*

This more complicated problem illustrates the care with which one must take limits. We consider a quantum mechanical system that is defined over the entire real axis and reduces to the half-harmonic oscillator in a particular limit. Our system is described by the superpotential

$$W(x,\zeta) = \zeta\ \Theta\left(-x\right) - \omega\,x\,\Theta\left(x\right) \tag{7.38}$$

where $\Theta\left(x\right)$ is the Heavyside step function. For finite values of ζ the system clearly has unbroken SUSY, as we see in Figure 7.3. We will show that SUSY must break down once the limiting procedure is carried out.

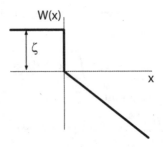

Fig. 7.3   The system has unbroken SUSY for finite values of *zeta* with $E_0^{(+)} \equiv \epsilon^+ = 0$.

We will do this by showing that the spectrum does not allow for a zero energy ground state and hence the supersymmetry is broken. The partner potentials $V_\pm$ are given by

$$V_\pm = \zeta^2\,\Theta\left(-x\right) + \left(\omega^2 x^2 \mp \omega\right)\Theta\left(x\right) \mp \zeta\,\delta(x)\,. \qquad (7.39)$$

For the V_+ potential, the solution on either side of the origin is:

$$x > 0 \qquad \psi(x) = A_1 D\left(\frac{\epsilon^+}{\omega}\,,\sqrt{2\omega}\,x\right)$$

$$x < 0 \qquad \psi(x) = A_2 \exp\left(\sqrt{\zeta^2 - 2\epsilon^+}\,x\right) \qquad (7.40)$$

where $D\left(\frac{\epsilon^+}{\omega},\sqrt{2\omega}\,x\right)$ is the parabolic cylindrical function[5]. To determine the eigenvalues, consider the following boundary conditions:

$$\psi(0_-) = \psi(0_+) \qquad (7.41)$$

and

$$\psi'(0_+) - \psi'(0_-) = -\zeta\psi(0_-)\,, \qquad (7.42)$$

which lead to the following two equations:

$$A_1 \frac{2^{\frac{\epsilon^+}{2\omega}}\sqrt{\pi}}{\Gamma\left(\frac{1}{2} - \frac{\epsilon^+}{2\omega}\right)} = A_2\left[\exp\left(\sqrt{\zeta^2 - 2\epsilon^+}\,x\right)\right]_{\zeta\to\infty,\ x=0_-} \qquad (7.43)$$

and

$$-2A_1\frac{2^{\frac{\epsilon^+}{2\omega}}\sqrt{\pi}\,\sqrt{\omega}}{\Gamma\left(-\frac{\epsilon^+}{2\omega}\right)} = A_2\left[\left(\sqrt{\zeta^2 - 2\epsilon^+} - \zeta\right)\exp\left(\sqrt{\zeta^2 - 2\epsilon^+}\,x\right)\right]_{\zeta\to\infty,\ x=0_-}$$

$$(7.44)$$

---

[5] See for example M. Abramowitz and I. A. Stegun, *Handbook of Mathematical Functions*, National Bureau of Standards Applied Mathematics Series - 55.

where

$$D\left(\frac{\epsilon}{\omega}, 0\right) = \frac{2^{\frac{\epsilon}{2\omega}}\sqrt{\pi}}{\Gamma\left(\frac{1}{2} - \frac{\epsilon}{2\omega}\right)} \quad \text{and} \quad D'\left(\frac{\epsilon}{\omega}, 0\right) = -2\frac{2^{\frac{\epsilon}{2\omega}}\sqrt{\pi}\sqrt{\omega}}{\Gamma\left(-\frac{\epsilon}{2\omega}\right)}.$$

The eigenvalues of the system are now determined by imposing the boundary conditions given by Eqs. (7.43) and (7.44). It is quite common for people to work with the ratio of the above two conditions; i.e., the logarithmic derivative. However, any such use of the logarithmic derivative should be done very carefully. In case of any confusion, one should revert back to the original conditions: (7.43) and (7.44). For example, since $\Gamma\left(-\frac{\epsilon^+}{2\omega}\right)\Big|_{\epsilon^+ \to 0} \to \infty$, both sides of Eq. (7.44) vanish for $\epsilon^+ = 0$, and hence each side of the ratio of Eq. (7.44) to Eq. (7.43) will also vanish for $\epsilon^+ = 0$ independent of whether or not $\epsilon^+ = 0$ is a solution of Eq. (7.43). We shall show that as $\zeta \to \infty$, $\epsilon^+ = 0$ is not a solution of Eq. (7.43), and hence not an eigenstate of the system. If these two fundamental equations (7.43 and 7.44) are satisfied, their ratios, in either order, should be satisfied as well. If either ratio reveals a contradiction, then they both become suspect, and one must question any conclusions derived from them, and return to the original boundary conditions, Eqs. (7.43) and (7.44).

Let us consider the "inverse logarithmic derivative": the ratio obtained by dividing Eq. (7.43) by Eq. (7.44). For $\epsilon^+ << \zeta^2$, it leads to

$$\frac{\sqrt{\omega}\ \Gamma\left(1 - \frac{\epsilon^+}{2\omega}\right)}{\epsilon^+\ \Gamma\left(\frac{1}{2} - \frac{\epsilon^+}{2\omega}\right)} = -\frac{\zeta}{\epsilon^+}. \tag{7.45}$$

For $\epsilon^+ \approx 0$, the above equation reduces to

$$\frac{\sqrt{\omega}}{\epsilon^+\sqrt{\pi}} = -\frac{\zeta}{\epsilon^+}.$$

Analyticity demands that both sides have the same degree of divergence. But the left hand side diverges as a simple pole $\frac{1}{\epsilon^+}$ while the right hand side diverges as $\zeta \cdot \frac{1}{\epsilon^+}$; i.e., only for finite ζ can both sides have the same divergence structure. In the limit $\zeta \to \infty$ the two sides do not diverge in the same way. Thus, the inverse logarithmic derivative has revealed a contradiction, casting both quotients $\frac{\psi'}{\psi}$ and $\frac{\psi}{\psi'}$ into doubt. We therefore return to the original boundary conditions (7.43) and (7.44) and examine their behavior at $x = 0$.

In the $\zeta \to \infty$-limit, both left (lhs) and right hand sides (rhs) of Eq. (7.44) reduce to zero for $\epsilon^+ = 0$. Thus, for $\epsilon^+ = 0$, it follows that $\frac{\text{LHS}}{\alpha} = \frac{\text{RHS}}{\beta}$, where α and β are *arbitrary non-zero* quantities. This is why $\epsilon^+ = 0$

is a solution of the condition on logarithmic derivative, Eq. (7.45). But this is not sufficient. That is, we not only need to satisfy Eq. (7.44), we also need to make sure that Eq. (7.43) is satisfied, for α and β the left and right hand sides respectively of that equation. And, we need to see if they are indeed nonzero. The logarithmic derivative fails to do this: it is satisfied for $\epsilon^+ = 0$ (because both sides of Eq. (7.44) reduce to zero), but without satisfying the boundary conditions on the wave function, Eq. (7.43), and without determining whether its value is nonzero at the origin.

Let us therefore carefully examine Eqs. (7.43) and (7.44). We are led to ask which limit should be taken first on the right hand side of Eq. (7.43). We shall show that both choices lead to a non-physical result.

If we take the $\zeta \to \infty$ limit first, then the right hand side of Eq. (7.43) gives $\psi(0_-) = A_2\, e^{-\zeta|x|} \to 0$. If we take $x = 0$ first, then $\psi(0_-) = A_2$. For the first choice, Eq. (7.41) then requires that $\psi(0_+) = 0$. Thus, Eq. (7.40) requires either that $A_1 = 0$ or that $D\left(\frac{\epsilon^+}{\omega}, 0\right) = 0$. Choosing $A_1 = 0$ produces $\psi(x) = 0$ for $x > 0$; thus, in the $\zeta \to \infty$ limit, this gives $\psi(x) = 0$ for all x. This is a non-physical solution. The other choice, $D\left(\frac{\epsilon^+}{\omega}, 0\right) = 0$, does not have an $\epsilon^+ = 0$ solution.

If on the other hand, $\psi(0_-) = A_2$, and then we take the $\zeta \to \infty$ limit, A_2 must be equal to zero, since $\psi(x) \to 0$ as $\zeta \to \infty$ for all $x < 0$: this is the behavior of $\psi(x)$ as a barrier becomes infinite. Once again, therefore, the continuity of the wave function (Eq. 7.41) leads to either $\psi(x) = 0$ for all x, or to $D\left(\frac{\epsilon^+}{\omega}, 0\right) = 0$, which does not have an $\epsilon^+ = 0$ solution.

Thus, while Eq. (7.44) does have $\epsilon^+ = 0$ as a solution, Eq. (7.43) does not. The ratio Eq. (6)/Eq. (5), determined by dividing by $\psi(0)$ and hence implicitly assuming $\psi(0) \neq 0$, erroneously gives $\epsilon^+ = 0$ as a solution, because, as we have shown, it gives $\psi(0_-) = \psi(0_+) = 0$.

The root of the problem is that Eq. (7.44) is itself incorrect as $\zeta \to \infty$, because the derivative of the wave function becomes discontinuous. The contribution of $\zeta \to \infty$ to Eq. (7.44) is identical to the contribution to the derivative condition at the boundary when one takes $\Lambda \to \infty$ for a well of depth Λ to arrive at an infinitely deep potential well. In that case, the limit $\Lambda \to \infty$ renders the derivative discontinuous and one must rely upon the condition of continuity of the wave function. Note that their δ-function singularity comes from the derivative of the step function, and not as an effect of the wall rising to infinite height. Thus, we cannot use Eq. (7.44) at all to determine the eigenvalues of the system. Instead, we must use Eq. (7.43).

The correct eigenenergies of H_+ can be obtained from the original boundary condition on the wave function, Eq. (7.43). Since the rhs is zero in the limit $\zeta \to \infty$, the lhs must have $\Gamma\left(\frac{1}{2} - \frac{\epsilon^+}{2\omega}\right) \to \infty$. Therefore, $\left(\frac{1}{2} - \frac{\epsilon^+}{2\omega}\right) = -n$, $n = 0, 1, 2, \ldots$, giving

$$\epsilon_n^+ = (2n + 1)\omega$$

with lowest eigenenergy $\omega > 0$. The eigenfunctions for these energies must be only the odd parity states $\psi_{-,n}(x) = B_{-,n}e^{-\frac{1}{2}\omega x^2}H_{2n+1}(\sqrt{\omega}x)$, which are naturally zero at the origin. The even solutions $\psi_{+,n}(x) = B_{+,n}e^{-\frac{1}{2}\omega x^2}H_{2n}(\sqrt{\omega}x)$ become discontinuous at the origin in the $\zeta \to \infty$ limit, violating the boundary condition Eq. (7.41).

We have shown that attempts to repair the broken supersymmetry of the HHO by regularization do not succeed, because a ground state energy equal to zero produces a non-physical wave function. We have identified the sources of the problem in the use of the logarithmic derivative, which involves division by zero, and neglects the discontinuity of the wave function as the barrier at the origin becomes infinite.

This section provides, as we mentioned earlier, a rather complicated but crucial lesson of the necessity for great care in taking limits to assess whether supersymmetry is broken.

Chapter 8

Potential Algebra

Let us recall from Chapter 2, in our discussion of the hydrogen atom, that for the time-independent Schrödinger equation in spherical coordinates, the angular solutions are the same for all spherically symmetric $V(r)$; namely, the spherical harmonics $Y(\theta, \phi)$. This is the quantum mechanical version of angular momentum conservation for spherically symmetric potentials.

Y is the solution to

$$\frac{1}{Y(\theta, \phi)} \left[\frac{1}{\sin\theta} \frac{\partial}{\partial\theta} \left(\sin\theta \frac{\partial Y(\theta, \phi)}{\partial\theta} \right) + \frac{1}{\sin^2\theta} \left(\frac{\partial^2 Y(\theta, \phi)}{\partial\phi^2} \right) \right] = -l(l+1).$$

(8.1)

If we multiply by $\sin^2\theta$ and divide by $Y(\theta, \phi) \equiv \Theta(\theta)\Phi(\phi)$, we separate the angular parts:

$$\frac{1}{\Theta} \left[\sin\theta \frac{\partial}{\partial\theta} \left(\sin\theta \frac{\partial}{\partial\theta}\Theta \right) + \left(-l(l+1)\sin^2\theta + \frac{1}{\Phi}\frac{\partial^2\Phi}{\partial\phi^2} \right) \right] = 0. \qquad (8.2)$$

Then the θ-dependent and ϕ-dependent terms must be separately constant. We thus define

$$\frac{1}{\Theta} \left[\sin\theta \frac{\partial}{\partial\theta} \left(\sin\theta \frac{\partial}{\partial\theta}\Theta \right) + \left(-l(l+1)\sin^2\theta \right) \right] \equiv m^2 ,$$

whence

$$\left[\frac{1}{\Phi}\frac{\partial^2\Phi)}{\partial\phi^2} \right] \equiv -m^2 ,$$

whose solution is $e^{im\phi}$.

If we demand single-valuedness for $\Phi(\phi)$ over the range 0 to 2π then $m = 0, \pm1, \pm2...$ [1] Substituting this into the Θ equation yields a differential equation whose solutions are the associated Legendre functions $P_l^m(\theta)$,

---

[1] This demand, common to many quantum mechanics texts, is not as simple as it sounds. The physical probability density is $\|\Phi(\phi)\|^2$, for which m cancels out. There are better arguments, such as the requirement that $\frac{\partial\Phi}{\partial\phi}$ be continuous for all values of ϕ, in order that its second derivative exist as demanded by the Schrödinger equation.

themselves obtainable from the Legendre polynomials $P_l(\theta)$

$$P_l^m(x) \equiv (1 - x^2)^{|m|/2} \left(\frac{d}{dx} \right)^m P_l(x) .$$

$P_l(x)$ is itself generated by

$$P_l(x) = \frac{1}{2^l l!} \frac{d}{dx}^l (x^2 - 1)^l .$$

Note that these definitions require that m and l be integers. The spherical harmonics $Y_l^m(\theta, \phi) = \Theta(\theta)\Phi(\phi)$ have only integer values of m and l.

All of these arguments for integer quantum numbers are self-consistent. The question is, do they cover all possibilities? Is it possible to have an angular momentum described by non-integer m and l? We will examine this question by starting first with the standard definition for angular momentum: $\mathbf{L} \equiv \mathbf{r} \times \mathbf{p}$. We will then write out its components $\mathbf{r} \equiv (x_1, x_2, x_3)$ and $\mathbf{p} \equiv (p_1, p_2, p_3)$, where $p_j \equiv -i\frac{\partial}{\partial x_j}$. (As usual, we take $\hbar = 1$.) From the commutation relations $[x_j, p_k] = i\delta_{jk}$, and the commutator theorem $[AB, C] = A[B, C] + [A, C]B$ [2] we can obtain commutation relations between the components (L_1, L_2, L_3) of $\mathbf{L}$.

Problem 8.1. *From the components of $\mathbf{L} \equiv \mathbf{r} \times \mathbf{p}$, show that*

(1) for i, j, k cyclic indices, $[L_i, x_i] = 0$, $[L_i, x_j] = ix_k$, $[L_i, x_k] = -ix_j$,
  $[L_i, p_i] = 0$, $[L_i, p_j] = ip_k$, and $[L_i, p_k] = -ip_j$.
(2) $[L_i, L_j] = iL_k$ for i, j, k cyclic indices.
(3) $[L^2, L_i] = 0$.

Since L^2 commutes with each of its components, the two have the same set of eigenfunctions. But since the components do not commute with each other, they cannot have the same set of eigenfunctions. We can thus choose only one of the components at a time and examine its relationship to L^2. As is customary, we will select L_3. If we call the common eigenfunctions f, then $L^2 f = \lambda f$ and $L_3 f = \mu f$. (The choice of these particular Greek letters is not arbitrary: as you might have guessed, λ will be related to l and μ to m.)

Now recall that in Chapter 4 we defined raising and lowering operators $A^\pm$ from which we were able to obtain the energy eigenvalues of the hamiltonian H. We shall use the same technique here, to obtain the respective eigenvalues λ and μ of L^2 and L_3, and relate them to each other. We define

---

[2] This can be proved in a straightforward fashion by writing out the commutators.

raising and lowering operators $L_\pm \equiv L_1 \pm iL_2$. We of course need to show that these are indeed raising and lowering operators, as we did for $A^\pm$. Recall that in that case, A^+ and A^- did not commute with each other: $[A^-, A^+] = 1$, nor with H: $[H, A^\pm] = \pm\omega A^\pm$. Indeed that is precisely what allowed us to build the ladder.

In the present case, L^2 will not do to build a ladder, since it commutes with each of its components, and therefore with any combination of them. We are then left with examining the commutators of $L_\pm$ with L_3.

$$[L_3, L_+] = [L_3, L_1] + i[L_3, L_2]$$

$$= iL_2 + i(-i)L_1$$

$$= iL_2 + L_1 = L_+.$$

Similarly, $[L_3, L_-] = -L_-$. We refer to this as "closing the algebra," since their commutators generate each other and no other operators.

Then for any eigenfunction of L_3,

$$L_+ f = (L_3 L_+ - L_+ L_3)f = (L_3 L_+ f - \mu L_+ f).$$

$$L_3(L_+ f) = (\mu + 1)(L_+ f).$$

So $L_+ f$ is an eigenfunction of L_3 with eigenvalue $\mu + 1$. Similarly, $L_- f$ is an eigenfunction of L_3 with eigenvalue $\mu - 1$. We have thus built a ladder of eigenfunctions and eigenvalues for L_3.

Now $\langle L^2 \rangle = \langle L_1^2 \rangle + \langle L_2^2 \rangle + \langle L_3^2 \rangle$ so $\langle L_3^2 \rangle \leq \langle L^2 \rangle$. This means that there is a $\mu_{\max}$. Similarly there exists a $\mu_{\min}$. That is, we have a ladder bounded at top and bottom. If we call f_{top} the eigenfunction corresponding to $\mu_{\max}$, and f_{bottom} the eigenfunction corresponding to $\mu_{\min}$, then "bounded" means $L_+ f_{\text{top}} = 0$ and $L_- f_{\text{bottom}} = 0$. We will use these relationships to obtain λ.

$$L_- L_+ = (L_1 - iL_2)(L_1 + iL_2) = L_1^2 + L_2^2 + i(L_1 L_2 - L_2 L_1) = L_1^2 + L_2^2 - L_3.$$

But $L_1^2 + L_2^2 = L^2 - L_3^2$. So

$$L_- L_+ = L^2 - L_3^2 - L_3.$$

Similarly

$$L_+ L_- = L^2 - L_3^2 + L_3.$$

We apply L_-L_+ to f_{top}, obtaining

$$0 = \left(L^2 f_{\text{top}} - L_3^2 - L_3\right) f_{\text{top}} = L^2 f_{\text{top}} - \mu_{\max}^2 f_{\text{top}} - \mu_{\max}^2 f_{\text{top}}$$

Collecting terms,

$$L^2 f_{\text{top}} = \mu_{\max}(\mu_{\max} + 1) f_{\text{top}} .$$

By tradition we define $l \equiv \mu_{\max}$, thus

$$L^2 f_{\text{top}} = l(l+1) f_{\text{top}} .$$

Similarly, operating on f_{bottom} with L_+L_- and defining $\bar{l} \equiv \mu_{\min}$,

$$L^2 f_{\text{bottom}} = \bar{l}(\bar{l} - 1) f_{\text{bottom}} .$$

But all functions f are degenerate with respect to L^2, since it played no role in the generation of the ladder. Thus $l(l+1) = \bar{l}(\bar{l} - 1)$, whose solutions are $l = \bar{l} - 1$ and $l + \bar{l} = 0$ The first cannot be, since by definition, $l > \bar{l}$, so $\bar{l} = -l$. This sets the limits on μ, which by tradition we now call m: $-l \leq m \leq l$. We have also seen that $L_\pm$ changes m by integer steps. Since we defined l and $\bar{l}(= -l)$ as the maximum and minimum values of m, the range of the ladder $-l \to l$ must be an integer $N = 2l$. So we now discover that $l = N/2$; i.e. l (and therefore m) can take on half-integer as well as integer values.

Let's review how this happened. We started with the traditional definition of angular momentum: $\mathbf{L} \equiv \mathbf{r} \times \mathbf{p}$. We used this to obtain the commutation relations between the components of $\mathbf{L}$. We then started with those commutation relations. Did they lead to $\mathbf{L} \equiv \mathbf{r} \times \mathbf{p}$ and only to that? We have now discovered that while they can of course do so: integer l, m, they also lead to another possibility as well: half-integer l, m. This possibility can be said to produce a new angular momentum, usually called spin, in analogy with Earth's "intrinsic" rotational angular momentum, while the traditional $\mathbf{L} \equiv \mathbf{r} \times \mathbf{p}$ had been analogous to its revolution to the sun. There is however an important difference. One can calculate Earth's rotation from its period and its radius. If you try this with the electron, it fails. Treating the electron as a classical solid sphere leads to its outer region moving faster than the speed of light. Furthermore, the electron's radius has thus far been found to be below any measurable value; it is presumed to be zero. Therefore the spin angular momentum must be a strictly quantum mechanical property intrinsic to the electron. In fact, spin is an intrinsic quantum mechanical property even for particles with finite radius.

Let us now return to the case of the traditional angular momentum, $\mathbf{L} = \mathbf{r} \times \mathbf{p}$, and investigate its hitherto abstract operators in their coordinate form, called their "realization." We do this for several reasons. One reason is to show that the eigenfunctions f are indeed what we expect from the analytic analysis of Chapter 2 summarized at the beginning of this chapter; viz., the spherical harmonics $Y_l^m(\theta, \phi)$. The second reason is to connect, somewhat surprisingly, the angular momentum equations with the Schrödinger equation for a particular potential.

We first write the angular momentum in its quantum mechanical form: $\mathbf{L} \equiv \mathbf{r} \times \frac{1}{i}\nabla$. Since we are working with spherically symmetric potentials, we use spherical polar coordinates.

$$\nabla = \hat{r}\frac{\partial}{\partial r} + \hat{\theta}\frac{1}{r}\frac{\partial}{\partial \theta} + \hat{\phi}\frac{1}{r\sin\theta}\frac{\partial}{\partial \phi}$$

Using the cross product relationships between the unit vector $\hat{r}$, $\hat{\theta}$, $\hat{\phi}$:

$$\hat{r} \times \hat{r} = 0, \ \hat{r} \times \hat{\theta} = \hat{\phi}, \ \hat{r} \times \hat{\phi} = -\hat{\theta},$$

we obtain

$$\mathbf{L} = \frac{1}{i}\left(\hat{\phi}\frac{\partial}{\partial \theta} - \hat{\theta}\frac{1}{\sin\theta}\frac{\partial}{\partial \phi}\right).$$

Now, however, we need the Cartesian components L_1, L_2, L_3 and the combinations $L_\pm \equiv L_1 \pm iL_2$ in terms of θ and ϕ. We use $\hat{\theta} = (\cos\theta\cos\phi)\hat{x_1} + (\cos\theta\sin\phi)\hat{x_2} - (\sin\theta)\hat{x_3}; \phi = -(\sin\phi)\hat{x_1} + (\cos\phi)\hat{x_2}$, and after collecting terms in $\hat{x_1}$, $\hat{x_2}$, $\hat{x_3}$, we obtain

$$L_1 = \frac{1}{i}\left(-\sin\phi\frac{\partial}{\partial \theta} - \cos\phi\cot\theta\frac{\partial}{\partial \phi}\right)$$

$$L_2 = \frac{1}{i}\left(\cos\phi\frac{\partial}{\partial \theta} - \sin\phi\cot\theta\frac{\partial}{\partial \phi}\right)$$

$$L_3 = \frac{1}{i}\frac{\partial}{\partial \phi}$$

$$L_\pm \equiv L_1 \pm iL_2 = \pm e^{\pm i\phi}\left(\frac{\partial}{\partial \theta} \pm i\cot\theta\frac{\partial}{\partial \phi}\right).$$

Writing the eigenfunction as $f = \Theta(\theta)\Phi(\phi)$, $L_3 f = mf$ yields $\Phi(\phi) = e^{im\phi}$. As we recall, $L^2 = L_+L_- + L_3^2 - L_3$ Applying the differential forms of these operators to f yields, after some algebra, the associated Legendre equation for Θ; i.e., Θ is the associated Legendre function P_l^m Therefore $f = Y_l^m(\theta, \phi)$, the spherical harmonic.

Problem 8.2.

(1) Apply L_- to Y_2^2 to obtain Y_2^1 up to a normalization constant.

(2) Apply the differential forms of $L+$, L_- and L_3 to Y_2^2.

(3) Show that Y_2^2 is an eigenfunction of L^2 with the correct eigenvalue.

Now as promised we will show that there is a way, using only the angular momentum commutation relations, to get a Schrödinger equation for a particular potential. In so doing, we will obtain the energy eigenvalues simply by use of these relations, rather than by the tedious brute-force analytic method. Note that this is analogous to the harmonic oscillator method, but run backwards. There, one knows the potential, uses it to define raising and lowering operators, and then obtains their commutation relations, from which the ladder of energies can be built. Here we will obtain a potential, knowing the commutation relations: $[L_i, L_j] = iL_k$ for i, j, k cyclic indices, and $[L^2, L_i] = 0$ or alternatively, by $L_\pm = L_1 \pm iL_2$: $[L_3, L_\pm] = \pm L_\pm$. and $[L^2, L_i] = 0$. We have seen that the raising and lowering operators are what we used to construct the ladder of states, as, for example in the above problem, so we suspect (correctly) that the latter form will be more useful.

We want a Schrödinger equation $H\psi = E\psi$; i.e. an eigenvalue equation. Then the operator we construct from the angular momentum operators must be the product of raising and lowering operators to insure that we obtain back to the same eigenfunction with which we started.

$$L_\pm \equiv L_1 \pm iL_2 = \pm e^{\pm\phi} \left[\frac{\partial}{\partial\theta} \pm \cot\theta \left(i\frac{\partial}{\partial\phi} \right) \right]. \qquad (8.3)$$

We also want an equation that can be cast as a Schrödinger equation; i.e.,with no first derivative.

We choose the product L_+L_-.

$$L_+L_- = - e^{i\phi} \left[\frac{\partial}{\partial\theta} + \cot\theta \left(i\frac{\partial}{\partial\phi} \right) \right] \cdot e^{-i\phi} \left[\frac{\partial}{\partial\theta} - \cot\theta \left(i\frac{\partial}{\partial\phi} \right) \right]$$

$$= - \left[\frac{\partial}{\partial\theta} + \cot\theta \left(i\frac{\partial}{\partial\phi} + 1 \right) \right] \left[\frac{\partial}{\partial\theta} - \cot\theta \left(i\frac{\partial}{\partial\phi} \right) \right]$$

$$= - \left[\frac{\partial^2}{\partial\theta^2} + \cot\theta \left(i\frac{\partial}{\partial\phi} + 1 \right) \frac{\partial}{\partial\theta} - \cot\theta \left(i\frac{\partial}{\partial\phi} \right) \frac{\partial}{\partial\theta} + \mathrm{cosec}^2\theta \left(i\frac{\partial}{\partial\phi} \right) \right.$$
$$\left. - \cot^2\theta \left(i\frac{\partial}{\partial\phi} + 1 \right) \left(i\frac{\partial}{\partial\phi} \right) \right]$$

$$= - \left[\frac{\partial^2}{\partial\theta^2} + \cot\theta \frac{\partial}{\partial\theta} - \cot^2\theta \left(i\frac{\partial}{\partial\phi} \right)^2 + \left(i\frac{\partial}{\partial\phi} \right) \right] \qquad (8.4)$$

Acting on an eigenfunction $\psi(\theta)e^{im\phi}$, the above operator gives:

$$L_+L_-\psi(\theta)e^{im\phi} = - \left[\frac{\partial^2}{\partial\theta^2} + \cot\theta\frac{\partial}{\partial\theta} - m^2\cot^2\theta + m\right]\psi(\theta)e^{im\phi}. \quad (8.5)$$

I.e., the action of the operator on $\psi(\theta)$ would be very similar to that of the Schrödinger operator only if we did not have the first order derivative $\cot\theta\frac{\partial}{\partial\theta}$.

We change variable to $z = z(\theta)$ where this functional dependence will be determined by demanding that we obtain a Schrödinger type operator in the new variable. This substitution gives $\frac{\partial}{\partial\theta} = \frac{\partial z}{\partial\theta}\frac{\partial}{\partial z}$ and $\frac{\partial^2}{\partial\theta^2} = \left(\frac{\partial z}{\partial\theta}\right)^2\frac{\partial^2}{\partial z^2} + \left(\frac{\partial^2 z}{\partial\theta^2}\right)\frac{\partial}{\partial z}$. Thus, terms $\frac{\partial^2}{\partial\theta^2} + \cot\theta\frac{\partial}{\partial\theta}$ go into

$$\frac{\partial^2}{\partial\theta^2} + \cot\theta\frac{\partial}{\partial\theta} = \left(\frac{\partial z}{\partial\theta}\right)^2\frac{\partial^2}{\partial z^2} + \left[\left(\frac{\partial^2 z}{\partial\theta^2}\right) + \cot\theta\frac{\partial z}{\partial\theta}\right]\frac{\partial}{\partial z}.$$

If we choose z such that the expression within square bracket vanishes, we will only have a second derivative term in the operator of Eq. (8.5). Solving the constraint $\frac{\partial^2 z}{\partial\theta^2} + \cot\theta\frac{\partial z}{\partial\theta} = 0$, we obtain $\frac{\partial z}{\partial\theta} = \operatorname{cosec}\theta$; i.e.,

$$z = \ln\left[\tan\left(\frac{\theta}{2}\right)\right]. \quad (8.6)$$

This change of variable maps the domain $(0, \pi)$ to $(-\infty, \infty)$: the domain of the one-dimensional Schrödinger equation. In these new coordinates L_+ and L_- become

$$L_\pm = e^{\pm i\phi}\left(-i\sinh z\frac{\partial}{\partial\phi} \pm \cosh z\frac{\partial}{\partial z}\right), \quad (8.7)$$

while L_3 remains unchanged. Like the A^+A^- operator we have seen before, we would like to see if we can associate the operator L_+L_- with a quantum mechanical hamiltonian. The eigenvalue equation $L_+L_-\psi = \eta\psi$ yields,

$$\left[-\sinh^2 z\frac{\partial^2}{\partial\phi^2} - i\frac{\partial}{\partial\phi} - \cosh^2 z\frac{\partial^2}{\partial z^2}\right]\psi(z,\phi) = \eta\psi(z,\phi). \quad (8.8)$$

Dividing this equation by $\cosh^2 z$, using $\operatorname{sech}^2 z = 1 - \tanh^2 z$, and rearranging the terms, we get

$$\left[-\frac{\partial^2}{\partial z^2} + \left(-i\frac{\partial}{\partial\phi} + \frac{\partial^2}{\partial\phi^2} - \eta\right)\operatorname{sech}^2 z\right]\psi(z,\phi) = \frac{\partial^2}{\partial\phi^2}\psi(z,\phi). \quad (8.9)$$

From Eq. (8.2), the eigenfunctions of the operator L_3 are given by $\psi(z,\phi) = g(z)e^{im\phi}$. Then Eq. (8.9) yields

$$\left[-\frac{\partial^2}{\partial z^2} + (m - m^2 - \eta)\operatorname{sech}^2 z\right]g(z) = -m^2 g(z). \quad (8.10)$$

Note that this is now a one-dimensional Schrödinger equation. The m's are "encoding" the behavior of $\frac{\partial}{\partial \phi}$'s on the eigenfunctions of L_3, $e^{i m \phi}$. Note that this is the typical Schrödinger equation problem run backwards. In the former we know the behavior of the operator $\frac{\partial}{\partial \phi}$, and obtain quantum the numbers m. Here we know the m's and "decode" them to retrodict $\frac{\partial}{\partial \phi}$.

Since $L_+ L_- = L^2 - L_3^2 + L_3$, the eigenvalue η from Eq. (8.8) is given by $\eta = l(l+1) - m^2 + m$. Substituting this expression in Eq. (8.10) leads to the following Schrödinger equation

$$\left[-\frac{\partial^2}{\partial z^2} - l(l+1) \operatorname{sech}^2 z \right] g(z) = -m^2 g(z) \qquad (8.11)$$

where $V(z, m, l) = -l(l+1) \operatorname{sech}^2 z$ is the potential and $-m^2$ is the corresponding energy.

8.1   Algebra of Shape Invariance

We have seen in the previous section that starting from the commutation relations for the angular momentum operators we arrive at a Schrödinger equation for the sech^2 potential. Therefore, we learned that there is a connection between the algebra of angular momentum operators and the sech^2 shape invariant potential. Can we generalize this connection to all shape invariant potentials? In this section we will further explore this question. More precisely, we will explore wether there is an algebra of some sort of generalized angular momentum-like operators that corresponds to each shape invariant potential. The complete mathematical apparatus needed to fully answer this question is beyond the scope of our book[3]; however, we can cannot leave the chapter without emphasizing some beautiful results concerning this subject.

Let us start by observing that the shape invariance condition (6.2) written in terms of A^+ and A^-

$$A^+(z, a_0) A^-(z, a_0) - A^-(z, a_1) A^+(z, a_1) = R(a_0) , \qquad (8.12)$$

is *almost* a commutation relation, if A^+ and A^- would have the same arguments. This suggests that we could use A^+ and A^- to build some new operators, which we denote by J_+, J_- and J_3, which in analogy with their angular momentum counterparts L_+, L_- and L_3, satisfy similar commutations relations

$$[J_3, J_+] = J_+ , \quad [J_3, J_-] = -J_- , \quad [J_+, J_-] = f(J_3) . \qquad (8.13)$$

[3]A. Gangopadhyaya, J. Mallow, and U. Sukhatme, "Translational shape invariance and the inherent potential algebra", *Phys. Rev. A* **58**, 4287 (1998).

When $f(J_3) = 2J_3$, we recover the usual angular momentum algebra case. In general, we design the operators $J_\pm$ and J_3 such that $f(J_3)$ in Eq. (8.13) is to be connected to $R(a_0)$ and therefore with the shape invariance condition (8.12). Our new algebra will be designed in such a way, that it will encode in its third commutation relation the analog of the shape invariance formula (8.12).

Using Eq. (8.12) for guidance, we see that it would resemble the third equation in (8.13) if somehow we make the connection $a_0 \propto J_3$. Let us define

$$J_+ = e^{is\phi} \mathcal{A}^+ \ , \quad J_- \equiv (J_+)^\dagger = \mathcal{A}^- e^{-is\phi} \tag{8.14}$$

where s is a constant parameter. The operator $\mathcal{A}^-$ is obtained from $A^- \equiv A^-(z, a_0)$ by introducing an auxiliary variable ϕ independent of z and replacing the parameter a_0 by a function $\chi(i\partial_\phi)$

$$A^- \equiv A^-(z, a_0) \xrightarrow{a_0 \mapsto \chi(i\partial_\phi)} \mathcal{A}^- \equiv \mathcal{A}^-(z, \chi(i\partial_\phi)) \ . \tag{8.15}$$

Similarly $\mathcal{A}^+ = (\mathcal{A}^-)^\dagger$. The reason for the substitution (8.15) will become evident after this short calculation

$$[J_+ \, , J_-] = J_+ J_- - J_- J_+ \tag{8.16}$$
$$= e^{is\phi} \mathcal{A}^+(z, \chi(i\partial_\phi)) \, \mathcal{A}^-(z, \chi(i\partial_\phi)) \, e^{-is\phi} - \mathcal{A}^-(z, \chi(i\partial_\phi)) \, \mathcal{A}^+(z, \chi(i\partial_\phi))$$
$$= \mathcal{A}^+(z, \chi(i\partial_\phi + s)) \, \mathcal{A}^-(z, \chi(i\partial_\phi + s)) - \mathcal{A}^-(z, \chi(i\partial_\phi)) \, \mathcal{A}^+(z, \chi(i\partial_\phi)) \ .$$

In the last line, we have used the fact that the operator $\partial_\phi \phi$ acting on a arbitrary function $f(\phi)$ yields $\partial_\phi \phi \, f(\phi) = \phi(\partial_\phi + 1) \, f(\phi)$. This implies $(i\partial_\phi) \, e^{-is\phi} = e^{-is\phi} \, (i\partial_\phi + s)$.

Observe now that the right hand side of Eq. (8.16) is similar to the left hand side of Eq. (8.12) provided that we make the following mappings:

$$a_0 \mapsto \chi(i\partial_\phi) \quad , \quad a_1 \mapsto \chi(i\partial_\phi + s) \ . \tag{8.17}$$

The parameters of shape invariance a_0 and a_1, can always be connected via a function η such that $a_1 = \eta(a_0)$. For example in the case of the Pöschl-Teller potential (6.8) the change of parameters is a simple translation: $a_1 \equiv \eta(a_0) = a_0 + 1$ where $a_0 = b$. We should remember that the identification, Eq. (8.17) is valid as long as we judiciously choose the function $\chi(i\partial_\phi)$ such that

$$\chi(i\partial_\phi + s) = \eta(\chi(i\partial_\phi)) \ . \tag{8.18}$$

Equation (8.18) is a rather general, indeed generous constraint. In the above example, where the change of parameters was the simple translation, the function modeling this translation is the identity function $\chi(z) = z$. Indeed we have in general for translational shape invariance $\chi(z + s) = z + s = \chi(z) + s$, which models perfectly the desired change of parameters for the infinite well and $cosec^2$ partner potentials if $s = 1$.

Next, we define the operator J_3 such that $[J_3, J_\pm] = \pm J_\pm$. Since $\left[-\frac{i}{s}\,\partial_\phi, \mathrm{e}^{\pm is\phi}\right] = \pm\,\mathrm{e}^{\pm is\phi}$, by direct calculation we can check that

$$J_3 = k - \frac{i}{s}\,\partial_\phi \tag{8.19}$$

satisfies these requirements. Here k is an arbitrary constant.

The last step in constructing the algebra for shape invariant systems is to identify the function $F(J_3)$ from (8.13). Comparing Eq. (8.12) with its extended form Eq. (8.16), and writing Eq. (8.19) as $s\,k - s\,J_3 = i\partial_\phi$, we have

$$f(a_0) \mapsto f\left(\chi(i\partial_\phi)\right) = f\left(\chi(s\,k - sJ_3)\right) \equiv F(J_3)\ . \tag{8.20}$$

Putting together all of the above observations, we observe that in fact, we have proved the following theorem:

To any shape invariant system characterized by

$$A^+(z, a_1)A^-(z, a_1) - A^-(z, a_0)A^+(z, a_0) = f(a_0)\ ,\quad a_1 = \eta(a_0) \tag{8.21}$$

we can associate an algebra generated by

$$J_+ = \mathrm{e}^{is\phi}\,\mathcal{A}^+\left(z, \chi(i\partial_\phi)\right) \tag{8.22}$$

$$J_- = \mathcal{A}^-\left(z, \chi(i\partial_\phi)\right)\mathrm{e}^{-is\phi} \tag{8.23}$$

$$J_3 = k - \frac{i}{s}\,\partial_\phi \tag{8.24}$$

satisfying the commutation relations

$$[J_3, J_+] = J_+\ ,\quad [J_3, J_-] = -J_-\ ,\quad [J_+, J_-] = F(J_3)\ , \tag{8.25}$$

where

$$F(J_3) = f\left(\chi(s\,k - sJ_3)\right)\ . \tag{8.26}$$

The function f in Eq. (8.26) is given by the shape invariance condition (8.21), while the function χ must satisfy the compatibility equation: $\chi(i\partial_\phi + s) = \eta\left(\chi(i\partial_\phi)\right)$, where η models the change of parameter $a_1 = \eta(a_0)$ of Eq. (8.21).

As an example, let us find the algebra for the cyclic shape-invariant potentials.

8.1.1 *Cyclic Potentials*

Let us consider a particular change of parameters given by the following cycle (or chain):

$$a_0, \ a_1 = f(a_0), \ a_2 = f(a_1), \dots, \ a_k = f(a_{k-1}) = a_0, \qquad (8.27)$$

and choose $R(a_i) = a_i \equiv \omega_i$. This choice generates a particular class of potentials known as cyclic potentials[4].

Cyclic potentials form a series of shape invariant potentials; the series repeats after a cycle of k iterations. In Figure (8.1) we show the first potential $V(x, a_0)$ from a 3-chain ($k = 3$) of cyclic potentials, corresponding to $\omega_0 = 0.15, \omega_1 = 0.25, \omega_2 = 0.60$.

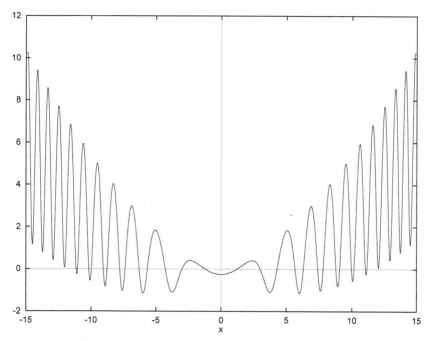

Fig. 8.1 First potential $V(x, a_0)$ from a 3-chain ($k = 3$).

Such potentials have an infinite number of periodically spaced eigenvalues. More precisely, the level spacings are given by $\omega_0, \omega_1, \dots, \omega_{k-1}, \omega_0, \omega_1, \dots, \omega_{k-1}, \omega_0, \omega_1, \dots$, as illustrated in Figure (8.2).

---

[4]U. Sukhatme, C. Rasinariu, A. Khare, "Cyclic Shape Invariant Potentials", *Phys. Lett.* **A 234**, 401-409 (1997).

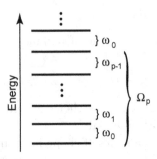

Fig. 8.2   Energy levels of a cyclic shape invariant potential of order p. The lowest level is at zero energy.

In order to generate the change of parameters (8.27) the function f should satisfy $f(f(\ldots f(x)\ldots)) \equiv f^k(x) = x$. The projective map

$$f(y) = \frac{\alpha y + \beta}{\gamma y + \delta} \quad , \tag{8.28}$$

with specific constraints on the parameters $\alpha, \beta, \gamma, \delta$, satisfies such a condition.

The next step is to identify the algebra behind this model. For this, we need to find the function χ satisfying the equation

$$\chi(z + p) = f(\chi(z)) \equiv \frac{\alpha \chi(z) + \beta}{\gamma \chi(z) + \delta} \quad . \tag{8.29}$$

It is a difference equation and its general solution is given by

$$\chi(z) = \frac{(\lambda_1 - \delta)\lambda_1^{z/p} + (\lambda_2 - \delta)\lambda_2^{z/p} B(z)}{\gamma \left[\lambda_1^{z/p} + \lambda_2^{z/p} B(z) \right]} \quad , \tag{8.30}$$

where $\lambda_{1,2}$ are solutions of the equation $(x - \alpha)(x - \delta) - \beta\gamma = 0$. For simplicity $B(z)$ can be chosen to be an arbitrary constant. Plugging this expression in Eqs. (8.25) we obtain:

$$[J_3, J_\pm] = \pm J_\pm \quad ;$$

$$[J_+, J_-] = -\frac{1}{c} \frac{A(\lambda_1 - \delta)\lambda_1^{-J_3} + B(\lambda_2 - \delta)\lambda_2^{-J_3}}{A\lambda_1^{-J_3} + B\lambda_2^{-J_3}} \quad . \tag{8.31}$$

We can see that this is a nonlinear algebra, which we expected from the formalism.

PART 2
Applications

Chapter 9

Special Functions and SUSYQM

An interesting application of shape invariance is the investigation of polynomial functions and their recursion relations[1]. The process is as follows: We start with the differential equation for a known polynomial. This will in general be a second order equation which contains a first order term. We need to recast the differential equation as a Schrödinger type equation in order to apply SUSYQM. That means we must eliminate the first-order term by a suitable transformation. Once this is done, SUSYQM does the rest.

Before we begin, however, let us remind ourselves of the shape invariance relations between adjacent wave functions, as we will need them in what follows.

$$\psi_{n+1}^{(-)}(x, a_0) = \frac{A^+(x, a_0)\psi_n^{(-)}(x, a_1)}{\sqrt{E_n^{(+)}(a_0)}} \tag{9.1}$$

and

$$\psi_n^{(-)}(x, a_1) = \frac{A^-(x, a_0)\psi_{n+1}^{(-)}(x, a_0)}{\sqrt{E_{n+1}^{(-)}(a_0)}}. \tag{9.2}$$

9.1 Hermite Polynomials

As an example, let us start with the Hermite differential equation:

$$\frac{d^2 H_n(\xi)}{d\xi^2} - 2\xi\frac{dH_n(\xi)}{d\xi} + 2nH_n(\xi) = 0 . \tag{9.3}$$

---

[1]The first studies of this method can be found in R. Dutt, A. Gangopadhyaya, and U. Sukhatme, "Noncentral Potentials And Spherical Harmonics Using Supersymmetry And Shape Invariance", *Am. J. Phys.* **65**, 400-403 (1997) and H. Rosu and J. R. Guzman, "Gegenbauer Polynomials And Supersymmetric Quantum Mechanics", *Nuovo Cimento della Societa Italiana di Fisica B.* **112**, 941-942 (1997).

.

We can eliminate the first-order term by choosing the ansatz $H_n(\xi) = f_n(\xi)\psi_n(\xi)$. Substituting this into the Hermite equation, we get:

$$\psi_n'' f_n + \psi_n' [2f_n' - 2f_n\xi] + \psi_n [f_n'' - 2f_n'\xi + f_n(2n)] = 0 . \qquad (9.4)$$

Setting the coefficient of the $\psi_n'(\xi)$ term equal to 0 gives $2f_n' - 2f_n\xi = 0$, which is solved by $f_n(\xi) = c_n e^{\frac{\xi^2}{2}}$. Replacing $f_n(\xi)$ in Eq. (9) and making the substitution $K_n \equiv 2n + 1$ yields:

$$-\psi_n'' + \xi^2 \psi_n = K_n \psi_n . \qquad (9.5)$$

This equation is just the Schrödinger equation for the harmonic oscillator, and the corresponding superpotential is $W(\xi) = a\xi$, where a is a constant. The SUSY partner potentials are given by $V_-(\xi, a) = a^2\xi^2 - a$ and $V_+(\xi, a) = a^2\xi^2 + a$. First, note that this set of potentials is shape invariant since

$$V_+(\xi, a) = V_-(\xi, a) + 2a .$$

(In standard texts, $\xi \equiv \sqrt{(\frac{2m\omega}{\hbar})}x$, which in our convention is $\sqrt{\frac{\omega}{2}}x$. Thus, $a = \sqrt{\frac{\omega}{2}}$.) Note that the parameter a has no subscript j: the shift is the same for all partner potentials. This is why the harmonic oscillator "collapses" into one ladder, as a special case of shape invariance.

Comparison with the potential term in Eq. (9.5) requires that we set $a = 1$. Now, the potential $V_-(\xi, a)$ differs from the traditional harmonic oscillator potential by an additive constant of -1. This is needed to make the SUSYQM ground state equal to 0.

The ground state eigenfunction can be determined using differential realization of $A^- |0\rangle = 0$.

$$\psi_0^{(-)} = e^{-\int \xi d\xi} = e^{-\frac{\xi^2}{2}} .$$

Using shape invariance, we obtain the higher states by successive application of $A^+(\xi, a_n)$, where $A^+ = \left[-\frac{d}{d\xi} + \xi \right]$.

We can now obtain recursion relations for the Hermite polynomials. Since $H_n(\xi) = f_n(\xi)\psi_n(\xi) = c_n e^{\frac{\xi^2}{2}} \psi_n(\xi)$, inverting this relation gives

$$\psi_n(\xi) = N_n H_n(\xi) e^{-\frac{\xi^2}{2}} , \qquad (9.6)$$

where $N_n = (\frac{1}{\pi})^{\frac{1}{4}} \frac{1}{\sqrt{2^n n!}}$ in keeping with the traditional normalization of the Hermite polynomials.

Applying the SUSY operators $A^+(\xi, a_n)$ and $A^-(\xi, a_n)$ leads to the following recursion relations:

$$H_n(\xi) = \left[2\xi - \frac{d}{d\xi} \right] H_{n-1}(\xi)$$

and

$$2nH_{n-1}(\xi) = \left[\frac{d}{d\xi}\right] H_n(\xi).$$

Problem 9.1. *Work out the details.*

These are two of the familiar Hermite polynomial recursion relations.

If you compare what we have done here with Chapter 2, you will see that in a sense we have run the harmonic oscillator scheme backwards. There, we began with the Schrödinger equation and obtained the wave functions as products of the decaying Gaussian functions and the Hermite polynomials. The recursion relations were not derived, although they might have been found by careful inspection. In the example here, we began with the Hermite Equation, turned it into a Schrödinger equation, then used SUSYQM and shape invariance to derive the recursion relations.

9.2 Associated Legendre Polynomial

This time we will start with the Schrödinger equation for a free particle and connect it to the associated Legendre polynomial. In three dimensions, the Schrodinger equations is given by

$$-\vec{\nabla}^2\psi \equiv -\frac{1}{r}\frac{\partial^2}{\partial r^2}(r\psi) - \frac{1}{r^2\sin\theta}\frac{\partial}{\partial\theta}\left(\sin\theta\frac{\partial\psi}{\partial\theta}\right) - \frac{1}{r^2\sin^2\theta}\frac{\partial^2\psi}{\partial\phi^2} = E\psi \quad.$$
(9.7)

Eq. (9.7) permits a solution via separation of variables, if one writes the wave function in the form

$$\psi(r,\theta,\phi) = R(r)P(\theta)e^{im\phi} \quad.$$
(9.8)

The equations satisfied by the quantities $R(r)$ and $P(\theta)$ are:

$$\frac{d^2R}{dr^2} + \frac{2}{r}\frac{dR}{dr} + \left(E - \frac{\ell(\ell+1)}{r^2}\right)R = 0 \quad,$$
(9.9)

$$\frac{d^2P_{\ell,m}}{d\theta^2} + \cot\theta\frac{dP_{\ell,m}}{d\theta} + \left[\ell(\ell+1) - \frac{m^2}{\sin^2\theta}\right]P_{\ell,m} = 0.$$
(9.10)

Since Eq. (9.10) will be transformed into the associated Legendre differential equation, our object of interest, we will ignore Eq. (9.9) for R. In addition, we note that we explicitly exhibited the ℓ, m-dependence in Eq. (9.10) as these parameters will become the indices for the associated Legendre polynomials P. To solve this equation by the SUSYQM method, we need to first re-cast it into a one-dimensional Schrödinger-like equation. In order to do this we substitute $\theta = f(z)$, where $f(z)$ is an arbitrary function to be soon determined. This substitution yields

$$\frac{d^2 P_{\ell,m}}{dz^2} + \left[-\frac{f''}{f'} + f' \cot f \right] \frac{dP_{\ell,m}}{dz} + f'^2 \left[\ell(\ell+1) - \frac{m^2}{\sin^2 f} \right] P_{\ell,m} = 0 \ .$$

$$(9.11)$$

Problem 9.2. *Show that Eq. (9.10) leads to Eq. (9.11).*

In order for Eq. (9.11) to be a Schrödinger-like equation, the first derivative term in will need to vanish. This requires

$$\frac{f''}{f'} = f' \cot f \ . \tag{9.12}$$

Solving the above equation, we obtain

$$\theta \equiv f = 2 \tan^{-1}(e^z) \qquad \text{or} \qquad e^z = \tan \frac{\theta}{2}. \tag{9.13}$$

The range of the new variable z is $-\infty < z < \infty$. From Eq. (9.13), we find $\sin \theta = \text{sech} z$ and $\cos \theta = -\tanh z$. Eq. (9.10) now reads

$$-\frac{d^2 P_{\ell,m}}{dz^2} + \underbrace{\left(\ell^2 - \ell(\ell+1) \, \text{sech}^2 z \right)}_{\text{Pöschl-Teller Potential}} P_{\ell,m} = \left(\ell^2 - m^2 \right) P_{\ell,m} \ . \tag{9.14}$$

As we have seen in Chapter 6, this is the Schrödinger equation for the Pöschl-Teller potential $V_-(z, \ell)$ generated by the superpotential $W(z, \ell) = \ell \tanh z$. Corresponding partner potentials $V_\pm(z, \ell)$ are $\ell^2 - \ell(\ell \mp 1) \, \text{sech}^2 z$, and the shape invariance condition relating them is

$$V_+(z, \ell) = V_-(z, \ell - 1) + \ell^2 - (\ell - 1)^2 \ .$$

Since the shape invariance condition reduces the value of ℓ in each step, and since ℓ must be positive for $V_-(z, \ell)$ to have the zero-energy ground state, this potential can only support a finite number of bound states. The eigenvalues are

$$E_n = \ell^2 - (\ell - n)^2 \ ; \qquad n = 0, 1, 2, \ldots N \ , \tag{9.15}$$

where N is the number of bound states this potential holds, and is equal to the largest integer contained in ℓ. The eigenfunctions $\psi_n(z, \ell)$ are obtained by using supersymmetry operators :

$$\psi_n(z; \ell) \propto A^+(z; \ell) A^+(z; \ell - 1) \cdots A^+(z; \ell - n + 1) \psi_0(z; \ell - n) \ , \tag{9.16}$$

where $A^+(z; \ell) \equiv \left(-\frac{d}{dz} + \ell \tanh z\right)$. The ground state wave function is $\psi_0(z; \ell) \sim (\operatorname{sech} z)^\ell$. From Eq. (9.14), we have $E_n = \ell^2 - m^2$. A quick comparison with Eq. (9.15) yields

$$n = \ell - m.$$

The corresponding eigenfunctions $\psi_{\ell-m}(z; \ell)$ are, from Eq. (9.10), the associated Legendre polynomials of degree ℓ, and are denoted by $P_{\ell,m}(\tanh z)$. Now that these $P_{\ell,m}(\tanh z)$ functions can be viewed as solutions of a Schrödinger equation, we can apply all the machinery one uses for a quantum mechanical problem. For example, we know that the parity of the n-th eigenfunction of a symmetric potential is given by $(-1)^n$, we readily deduce the parity of $P_{\ell,m}$ to be $(-1)^{\ell-m} = (-1)^{\ell+m}$. Also, the application of supersymmetry algebra results in identities that are either not very well known or not easily available. With repeated application of the A^+ operators, we can determine $P_{\ell,m}$ for a fixed value $(\ell - m)$. As an illustration, we explicitly work out all the polynomials for $\ell - m = 2$. The lowest polynomial corresponds to $\ell = 2, m = 0$. For a general ℓ, using Eq. (9.16), one gets

$$
\begin{aligned}
P_{\ell,\ell-2}(\tanh z) &\sim \psi_2(z; \ell) \\
&\sim A^+(z; \ell) A^+(z; \ell - 1) \psi_0(z; \ell - 2) \\
&\sim \left(-\frac{d}{dz} + \ell \tanh z\right)\left(-\frac{d}{dz} + (\ell - 1) \tanh z\right) \operatorname{sech}^{\ell-2} z \\
&\sim \left(-1 + (2\ell - 1) \tanh^2 z\right) \operatorname{sech}^{\ell-2} z \ .
\end{aligned}
\tag{9.17}
$$

I.e.,

$$
P_{\ell,\ell-2}(x) \sim \left(-1 + (2\ell - 1)x^2\right)\left(1 - x^2\right)^{\frac{\ell-2}{2}} \ .
$$

A similar procedure is readily applicable for other values of n. When one converts back to the original variable θ $(\operatorname{sech} z = \sin \theta \ ; \tanh z = -\cos \theta)$, the results are:

$$P_{\ell,\ell}(\cos \theta) \sim \sin^\ell \theta \tag{9.18}$$

$$P_{\ell,\ell-1}(\cos \theta) \sim \sin^{\ell-1} \theta \cos \theta$$

$$P_{\ell,\ell-2}(\cos \theta) \sim \left[-1 + (2\ell - 1)\cos^2 \theta\right] \sin^{\ell-2} \theta$$

$$P_{\ell,\ell-3}(\cos \theta) \sim \left[-3 + (2\ell - 1)\cos^2 \theta\right] \sin^{\ell-3} \theta \cos \theta$$

$$P_{\ell,\ell-4}(\cos \theta) \sim \left[3 - 6(2\ell - 3)\cos^2 \theta + (2\ell - 3)(2\ell - 1)\cos^4 \theta\right] \sin^{\ell-4} \theta \ .$$

Likewise, there are several recursion relations which are easily obtainable via SUSYQM methods. In particular, by applying A^- or A^+ once, we generate recursion relations of varying degrees (differing in the value of ℓ). From $\psi_{\ell-m}(z, \ell) = A^+(z, \ell)\psi_{\ell-m-1}(z, \ell - 1)$, we get

$$P_{\ell,m}(x) = \left((1 - x^2)\frac{d}{dx} + \ell x\right) P_{\ell-1,m} , \qquad (9.19)$$

and from $\psi_{\ell-m}(z, \ell) = A^-(z, \ell - 1)\psi_{\ell-m+1}(z, \ell + 1)$, we obtain

$$P_{\ell,m}(x) = \left((1 - x^2)\frac{d}{dx} + (\ell - 1) x\right) P_{\ell+1,m}. \qquad (9.20)$$

9.3   Laguerre Polynomials

We now consider the associated Laguerre polynomials, which as we mentioned earlier are the radial parts of the hydrogen atom wave functions. These polynomials are solutions of the equation:

$$x\frac{d^2 L_n^\alpha(x)}{dx^2} + (\alpha + 1 - x)\frac{dL_n^\alpha(x)}{dx} + nL_n^\alpha(x) = 0 . \qquad (9.21)$$

Here α is a constant parameter. To cast the above differential equation as a Schrödinger-like equation, we choose the ansatz $L_n^\alpha(x) = f_n^\alpha(x)U_n^\alpha(x)$. Elimination of the first derivative term leads to

$$f_n^\alpha(x) = c_n^\alpha x^{-\frac{(\alpha+1)}{2}} e^{\frac{x}{2}} .$$

Problem 9.3. *Work out the details.*

This substitution gives

$$-U'' + \left[\frac{(\alpha + 1)(\alpha - 1)}{4x^2} - \frac{(2n + \alpha + 1)}{2x} + \frac{1}{4}\right] U = 0 . \qquad (9.22)$$

As we shall see in Chapter 12, the coefficient of $\frac{1}{x^2}$; viz., $\frac{(\alpha+1)(\alpha-1)}{4}$ has to be larger than $-\frac{1}{4}$, otherwise the wave function would diverge too strongly at the origin. The spectrum would not be bounded from below; i.e., there could be no $E = 0$ ground state. This constrains the coefficient α to be real.

Equation (9.22) can be generated from the superpotential

$$W(x, n, \alpha) = \frac{2n + \alpha + 1}{2(\alpha + 1)} - \frac{(\alpha + 1)}{2x} .$$

The SUSYQM operators are given by $A = \left(\frac{d}{dx} + \frac{(2n+\alpha+1)}{2(\alpha+1)} - \frac{\alpha+1}{2x}\right)$ and $A^+ = \left(-\frac{d}{dx} + \frac{(2n+\alpha+1)}{2(\alpha+1)} - \frac{\alpha+1}{2x}\right)$ and the partner potentials are given by:

$$V_-(x, n, \alpha) = W^2 - W'$$
$$= \frac{(\alpha + 1)(\alpha - 1)}{4x^2} - \frac{(2n + \alpha + 1)}{4x} + \left[\frac{(2n + \alpha + 1)}{2(\alpha + 1)}\right]^2 \quad (9.23)$$

$$V_+(x, n, \alpha) = W^2 + W'$$
$$= \frac{(\alpha + 1)(\alpha + 3)}{4x^2} - \frac{(2n + \alpha + 1)}{4x} + \left[\frac{(2n + \alpha + 1)}{2(\alpha + 1)}\right]^2 . \quad (9.24)$$

They are related by the following shape invariance condition:

$$V_+(x, n, \alpha) = V_-(x, n - 1, \alpha + 2)$$
$$+ \left[\left(\frac{2n + \alpha + 1}{2(\alpha + 1)}\right)^2 - \left(\frac{2n + \alpha + 1}{2(\alpha + 3)}\right)^2\right] . \quad (9.25)$$

The solutions $U_n^\alpha(x)$ of the Schrödinger equation are then related to each other via Eq.(9.1) and Eq.(9.2).

From our definition of U, we can write

$$U_n^\alpha(x) = K_n^\alpha x^{\frac{(\alpha+1)}{2}} e^{-\frac{x}{2}} L_n^\alpha(x) , \quad (9.26)$$

where K_n^α is a constant of normalization. To ensure that the $L_n^\alpha(x)$ have their traditional normalization, we choose

$$K_n^\alpha = \sqrt{\frac{n!}{(2n + \alpha + 1)[(n + \alpha)!]^3}} .$$

Substitution of U into Eq. (9.1) gives

$$U_n^\alpha(x) = \frac{1}{\sqrt{E_{n,\alpha}^{(-)}}} A^+(x, n, \alpha) U_{n-1}^{\alpha+2} .$$

A tedious but straightforward simplification yields the following recursion relation among Laguerre polynomials:

$$L_n^a(x) = -\frac{(\alpha + 1)}{n(n + \alpha + 1)^2} \left[\left(\frac{n + \alpha + 1}{\alpha + 1}\right) x - (\alpha + 2) - x\frac{d}{dx}\right] L_{n-1}^{\alpha+2}(x) . \quad (9.27)$$

An additional recursion relation is obtained via the lowering operation A, Eq. (9.2): $U_{n-1}^{\alpha+2}(x) = \frac{1}{\sqrt{E_{n,a}^{(-)}}} A(x, n, \alpha) U_n^\alpha(x)$. This gives

$$L_{n-1}^{\alpha+2}(x) = -(n + \alpha + 1)\left[\frac{n}{x} + \frac{(\alpha + 1)}{x}\frac{d}{dx}\right] L_n^\alpha(x) . \quad (9.28)$$

These examples: the Hermite, the associated Legendre and the associated Laguerre polynomials, show the power of SUSYQM with shape invariance: they provide an algorithm for *deriving* recursion relations for polynomial functions. The technique can be applied to other sets of functions, provided that their differential equations can be recast as Schrödinger-like equations, whose superpotentials can then be determined.

Chapter 10

Isospectral Deformations

Given any one-dimensional potential with n bound states, one can ask the following question: *Are there any other potentials that have the same spectrum and the same reflection and transmission coefficients as the original potential?* This question is motivated by practical applications. For example, in the case of $\alpha - \alpha$ scattering it was observed that there are ambiguities involved in determining the potential even when the phase shifts and bound states are precisely known.

In this chapter we will show how we can use SUSYQM to build a family of new potentials $\hat{V}_-(\lambda; x)$ that have the same spectrum and the same reflection and transmission coefficients as the "original" potential $V_-(x)$. We will also show that the family of potentials $\hat{V}_-(\lambda; x)$ is unique. That means that starting from $\hat{V}_-(\lambda; x)$ and using the same technique, we do not arrive at a new family of isospectral potentials. The potential family

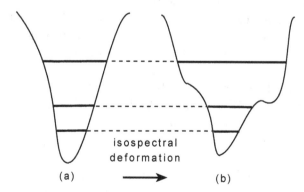

Fig. 10.1 Isospectral deformation. The potentials in (a) and (b) have the same spectrum and the same reflection and transmission coefficients.

described by $\hat{V}_-(\lambda; x)$ is known as the *isospectral deformation* of $V_-(x)$. Figure 10.1 is a sketch of a potential and its deformed partner.

10.1 One Parameter Isospectral Deformations

In the previous chapters we have seen that for a given potential $V_- = W^2 - W'$ its supersymmetric partner potential $V_+ = W^2 + W'$ has the same energy spectrum with the exception of the ground state $E_0^{(-)}$. We have

$$E_0^{(-)} = 0 \qquad\qquad\qquad E_{n+1}^{(-)} = E_n^{(+)} \quad \text{(isospectrality)}$$

$$\psi_n^{(+)} = \frac{d/dx + W(x)}{\sqrt{E_n^{(+)}}} \psi_{n+1}^{(-)} \qquad \psi_{n+1}^{(-)} = \frac{-d/dx + W(x)}{\sqrt{E_n^{(+)}}} \psi_n^{(+)} \quad (10.1)$$

where the superpotential $W(x)$ is given by

$$W(x) = -\psi_0^{(-)}(x)'/\psi_0^{(-)}(x) \equiv -\frac{d}{dx}\ln\psi_0^{(-)}(x) . \tag{10.2}$$

Also, as we shall see in Chapter 16, the reflection and transmission amplitudes for the partner potentials $V_\pm(X)$ are related by

$$r_-(k) = \frac{W_- + ik}{W_- - ik} r_+(k) , \qquad t_-(k) = \frac{W_+ - ik'}{W_- - ik} t_+(k) , \tag{10.3}$$

where $k = \sqrt{E - W_-^2}$, $k' = \sqrt{E - W_+^2}$, with $W_\pm = \lim_{x \to \pm\infty} W(x)$.

The almost isospectrality of $V_-(x)$ and $V_+(x)$ suggests that starting from $V_+(x)$ we can investigate whether the corresponding partner potential $V_-(x)$ is unique or not. We will find that there are other partner potentials $\hat{V}_-(\lambda, x)$ corresponding to a given $V_+(x)$. Consequently, we will prove that for a given potential $V_-(x)$ there exists a family of potentials $\hat{V}_-(\lambda, x)$ that all have the same spectrum. These are the isospectral deformations of $V_-(x)$.

In terms of the superpotential $W(x)$ we will assume that there exist other superpotentials $\hat{W}(x)$ which yield the same $V_+(x)$:

$$V_+(x) = \hat{W}^2(x) + \hat{W}'(x) , \tag{10.4}$$

or equivalently

$$W^2(x) + W'(x) = \hat{W}^2(x) + \hat{W}'(x) . \tag{10.5}$$

A trivial solution of Eq. (10.5) is $\hat{W} = W$. In general, if we set

$$\hat{W}(x) = W(x) + f(x) \tag{10.6}$$

and substitute it back into Eq. (10.5) we obtain

$$2\,W(x)f(x) + f(x)^2 + f'(x) = 0 \ . \tag{10.7}$$

Making the substitution $f(x) = 1/y(x)$ we obtain a simple differential equation

$$y'(x) = 1 + 2\,W(x)y(x) \tag{10.8}$$

that can be directly integrated. Its solution is

$$y(x) = \left(\int_{-\infty}^{x} e^{-2\int_{-\infty}^{t} W(u)du} dt + \lambda \right) e^{\int_{-\infty}^{x} 2\,W(t)dt} \ , \tag{10.9}$$

where λ is an integration constant. Now, we use Eq. (10.2) to rewrite (10.9) as

$$y(x) = \frac{\int_{-\infty}^{x} \left[\psi_0^{(-)}(t) \right]^2 dt + \lambda}{\left[\psi_0^{(-)}(x) \right]^2} \ . \tag{10.10}$$

Therefore,

$$f(x) \equiv 1/y(x) = \frac{\left[\psi_0^{(-)}(x) \right]^2}{\int_{-\infty}^{x} \left[\psi_0^{(-)}(t) \right]^2 dt + \lambda}$$

$$= \frac{d}{dx} \ln[I(x) + \lambda] \tag{10.11}$$

where $I(x) = \int_{-\infty}^{x} [\psi_0^{(-)}(t)]^2\, dt$. We conclude that the most general $\hat{W}(x)$ satisfying (10.5) is

$$\hat{W}(x) = W(x) + \frac{d}{dx} \ln[I(x) + \lambda] \ . \tag{10.12}$$

Finally, we are in a position to find the $\hat{V}_-(\lambda, x)$ corresponding to $\hat{W}(x)$. We find that all potentials given by

$$\hat{V}_-(\lambda, x) = \hat{W}(x)^2 - \hat{W}(x)' = V_-(x) - 2\frac{d^2}{dx^2} \ln[I(x) + \lambda] \tag{10.13}$$

have the same $V_+(x)$ and the same spectrum.

Problem 10.1. *Prove Eq. (10.13).*

Observe that when $\lambda \to \pm\infty$, $\hat{V}_-(\lambda, x) \to V_-(x)$. Therefore, $V_-(x)$ itself is a member of this one-parameter family.

Problem 10.2. *Using Eqs. (10.3) and (10.13), prove that the reflection and the transmission coefficients of $\hat{V}_-(\lambda, x)$ don't change: $\hat{r}_-(k) = r_-(k)$, $\hat{r}_-(k) = r_-(k)$.*

We have found that Eq. (10.13) generates an infinite family of isospectral potentials for $V_-(x)$. However the problem is not completely solved because we have not yet specified the acceptable range of values for the parameter λ. In order to see if there are any constrains for λ, we consider the ground state wave function $\hat{\psi}_0^{(-)}(x)$ corresponding to the deformed potential $\hat{V}_-(\lambda, x)$ and require that it be square integrable. Using Eqs. (10.12) and (10.2) we have successively:

$$\hat{\psi}_0^{(-)}(x) = N\, e^{-\int_{-\infty}^{x} \hat{W}(t)dt}$$

$$= N\, e^{-\int_{-\infty}^{x} \left(W(t) + \frac{d}{dt}\ln[I(t)+\lambda]\right)dt}$$

$$= N\, \frac{\psi_0^{(-)}(x)}{I(x) + \lambda}\,, \qquad (10.14)$$

where N is a normalization constant. After some calculations we find $N = \sqrt{\lambda(\lambda+1)}$. Therefore, the normalized ground state wave function for $\hat{V}_-(\lambda, x)$ is

$$\hat{\psi}_0^{(-)}(x) = \frac{\sqrt{\lambda(\lambda+1)}}{I(x) + \lambda}\, \psi_0^{(-)}(x)\,. \qquad (10.15)$$

Problem 10.3. *Calculate the normalization constant of Eq. (10.15).*

Note that $I(x)$ is a bounded quantity constrained to the interval $(0, 1)$. Consequently, we have three different possibilities for the range of values of λ:

(1) $\lambda < -1$ or $\lambda > 0$. In this case $\lambda(\lambda+1) > 0$ and the wave function (10.15) is a well behaved function which is square integrable. Thus the spectrum of $\hat{V}_-(\lambda, x)$ is identical with the spectrum of $V_-(x)$. Further more, we can see that in this case all the members of the family behave the same for $x \to \pm\infty$. If we know all the energy levels and the wave functions for $V_-(x)$ (i.e., if the potential is exactly solvable), then $\hat{V}_-(\lambda, x)$ is exactly solvable as well.

(2) $\lambda = -1$ or $\lambda = 0$. Because $I(x)$ vanishes at $x = \pm\infty$, in this case, the wave function $\hat{\psi}_0^{(-)}(x)$ is not square integrable. Thus SUSY is broken, and $\hat{V}_-(\lambda, x)$ loses a bound state. We will not explore this case any further. We simply mention that $\lambda = 0$ yields the Pursey potential and $\lambda = -1$ yields the Abraham-Moses potential. For more details the reader should consult the reference cited below[1].

(3) $-1 < \lambda < 0$. Because $I(x)$ is between 0 and 1, in this case, the potential $\hat{V}_-(\lambda, x)$ will be singular for certain values of x. These singularities raise significant questions about the physical meaning of such potentials, and we will not study them any further here.

10.2 Examples

As an example let us consider the Morse potential generated by the superpotential

$$W(x) = A - B e^{-\alpha x} , \quad A, B, \alpha > 0 . \tag{10.16}$$

Then, the potential is

$$V_-(x) = A^2 + B^2 e^{-2\alpha x} - 2B(A + \alpha/2)e^{-\alpha x}, \tag{10.17}$$

with energy levels

$$E_n = A^2 - (A - n\alpha)^2 , \quad n = 0, 1, 2, \ldots \tag{10.18}$$

and the corresponding wave functions given by

$$\psi_n = y^{s-n} e^{-y/2} L_n^{2s-2n}(y) ; \quad y = \frac{2B}{\alpha} e^{-\alpha x} , \quad s = A/\alpha , \tag{10.19}$$

where $L_n^k(x)$ are the usual Laguerre polynomials.

Now we have all the tools we need to construct the family of isospectral deformations of the Morse potential. Note that the calculations become a little tedious, but they present no conceptual difficulty. The reader is encouraged to use a symbolic algebra system such as Mathematica or Maple to speed up the computational process. We begin by calculating the ground state wave function. From Eq. (10.19) we obtain

$$\psi_0^{(-)}(x) = 2^{A/\alpha} e^{-\frac{Be^{-\alpha x}}{\alpha}} \left(\frac{Be^{-\alpha x}}{\alpha} \right)^{A/\alpha} . \tag{10.20}$$

---

[1] A. Khare and U. Sukhatme, "Phase equivalent potentials obtained from supersymmetry", *J. Phys. A* **22**, 2847 (1989).

Therefore we obtain

$$I(x) \equiv \int_{-\infty}^{x} \left[\psi_0^{(-)}(t)\right]^2 dt = 2^{\frac{2A}{\alpha}} \int_{-\infty}^{x} e^{-\frac{2Be^{-\alpha t}}{\alpha}} \left(\frac{Be^{-\alpha t}}{\alpha}\right)^{\frac{2A}{\alpha}} dt , \quad (10.21)$$

and consequently

$$\hat{V}_-(\lambda, x) = -\alpha B e^{-\alpha x} + \left(A - Be^{-\alpha x}\right)^2$$

$$-2 \left[-\frac{2^{\frac{4A}{\alpha}} e^{-\frac{4Be^{-\alpha x}}{\alpha}} \left(\frac{Be^{-\alpha x}}{\alpha}\right)^{\frac{4A}{\alpha}}}{\left(\lambda + 2^{\frac{2A}{\alpha}} \int_{-\infty}^{x} e^{-\frac{2Be^{-\alpha t}}{\alpha}} \left(\frac{Be^{-\alpha t}}{\alpha}\right)^{\frac{2A}{\alpha}} dt\right)^2} \right.$$

$$- \frac{2^{1+\frac{2A}{\alpha}} A B e^{-\frac{2Be^{-\alpha x}}{\alpha} - \alpha x} \left(\frac{Be^{-\alpha x}}{\alpha}\right)^{-1+\frac{2A}{\alpha}}}{\alpha \left(\lambda + 2^{\frac{2A}{\alpha}} \int_{-\infty}^{x} e^{-\frac{2Be^{-\alpha t}}{\alpha}} \left(\frac{Be^{-\alpha t}}{\alpha}\right)^{\frac{2A}{\alpha}} dt\right)}$$

$$\left. + \frac{2^{1+\frac{2A}{\alpha}} B e^{-\frac{2Be^{-\alpha x}}{\alpha} - \alpha x} \left(\frac{Be^{-\alpha x}}{\alpha}\right)^{\frac{2A}{\alpha}}}{\lambda + 2^{\frac{2A}{\alpha}} \int_{-\infty}^{x} e^{-\frac{2Be^{-\alpha t}}{\alpha}} \left(\frac{Be^{-\alpha t}}{\alpha}\right)^{\frac{2A}{\alpha}} dt} \right] . \quad (10.22)$$

Despite the apparent complexity of Eq. (10.22), we can recognize on its first line the usual Morse potential, and in the subsequent lines the terms characterizing the isospectral deformation. These terms vanish as $\lambda \to \pm\infty$ leaving us with the original Morse potential, as expected. Let us choose some straightforward numerical values for A, B, and α in order to simplify the calculations. This choice is useful in particular, as we want to visualize our results. Setting $A = \alpha = 1$ and $B = 2$ we obtain successively:

$$\psi_0^{(-)}(x) = 4e^{-2e^{-x}-x} \quad (10.23)$$

$$I(x) = 16e^{-4e^{-x}} \left(\frac{1}{16} + \frac{e^{-x}}{4}\right) \quad (10.24)$$

$$\hat{V}_-(\lambda, x) = -2e^{-x} + \left(1 - 2e^{-x}\right)^2 - 2 \left(-\frac{256e^{-8e^{-x}-4x}}{\left(16e^{-4e^{-x}} \left(\frac{1}{16} + \frac{e^{-x}}{4}\right) + \lambda\right)^2} \right.$$

$$\left. + \frac{64e^{-4e^{-x}-3x}}{16e^{-4e^{-x}} \left(\frac{1}{16} + \frac{e^{-x}}{4}\right) + \lambda} - \frac{32e^{-4e^{-x}-2x}}{16e^{-4e^{-x}} \left(\frac{1}{16} + \frac{e^{-x}}{4}\right) + \lambda} \right) . \quad (10.25)$$

In Figure 10.2 we illustrate the isospectral deformations of the Morse potential for three particular values of the λ parameter: $\lambda = 0.2$, $\lambda = 1.0$,

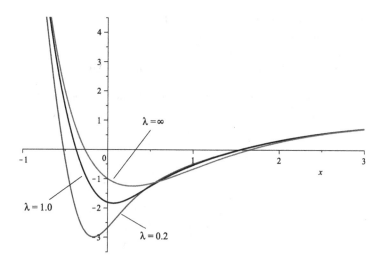

Fig. 10.2   Isospectral deformations of the Morse potential. The case $\lambda = \infty$ represents the original potential.

and $\lambda = \infty$. The last case corresponds to the initial, undeformed Morse potential.

Problem 10.4. *Construct the normalized ground state wave function corresponding to the deformed Morse potential and represent it graphically for $\lambda = 0.2$, $\lambda = 1.0$, and $\lambda = \infty$.*

Problem 10.5. *Repeat this procedure to construct the isospectral deformations and the corresponding ground state wave functions for the Coulomb potential.*

10.3   The Uniqueness of the Isospectral Deformation

Finally, we will address the question of the uniqueness of the isospectral deformation. If we start with any particular member $\hat{V}_-(\lambda, x)$ and again construct a one-parameter family of isospectral potentials, will this generate another distinct family? The answer is no, as we will see in the following proof.

For a potential $V_-(x)$, the one-parameter potential family is given by

$$\hat{V}_-(\lambda, x) = V_-(x) - 2\frac{d^2}{dx^2}\ln[I(x) + \lambda] \qquad (10.26)$$

where we take λ to be outside the $[-1, 0]$ interval. If we start with this $\hat{V}_-(\lambda, x)$ and deform it again, we obtain

$$\hat{\hat{V}}_- = \hat{V}_- - 2\frac{d^2}{dx^2}\ln[\hat{I} + \hat{\lambda}] = V_- - 2\frac{d^2}{dx^2}\ln[(I + \lambda)(\hat{I} + \hat{\lambda})] \qquad (10.27)$$

where for simplicity we dropped the explicit λ and x dependence. Let us estimate the value of $\hat{I}$. Using Eq. (10.15) we obtain successively

$$\hat{I}(x) = \int_{-\infty}^{x}\left[\hat{\psi}_0^{(-)}(t)\right]^2 dt$$

$$= \lambda(\lambda + 1)\int_{-\infty}^{x}\frac{\left[\psi_0^{(-)}(t)\right]^2}{[I(t) + \lambda]^2} dt$$

$$= \lambda(\lambda + 1)\int_{-\infty}^{x}\frac{\frac{d}{dt}I(t)}{[I(t) + \lambda]^2} dt$$

$$= \lambda(\lambda + 1)\left(\frac{1}{\lambda} - \frac{1}{I(x) + \lambda}\right). \qquad (10.28)$$

Consequently, Eq. (10.27) becomes

$$\hat{\hat{V}}_- = V_- - 2\frac{d^2}{dx^2}\ln\left\{(I(x) + \lambda)\left[\lambda(\lambda + 1)\left(\frac{1}{\lambda} - \frac{1}{I(x) + \lambda}\right) + \hat{\lambda}\right]\right\}$$

$$= V_- - 2\frac{d^2}{dx^2}\ln[I(x) + \mu], \qquad (10.29)$$

where

$$\mu = \frac{\hat{\lambda}\,\lambda}{\hat{\lambda} + \lambda + 1}. \qquad (10.30)$$

We observe that $\hat{\hat{V}}_-$ doesn't yield anything new. The only difference is that the deformation parameter λ was replaced by μ. One can prove that if both λ and $\hat{\lambda}$ are in the same range outside $[-1, 0]$ then μ is in the same range outside of this interval as well. Therefore, the one-parameter deformation yields a unique isospectral family.

Chapter 11

Generating Shape Invariant Potentials

In this chapter we will develop another method for generating shape invariant potentials. Most of the known exactly solvable superpotentials are such that their parameters differ by an additive constant; i.e., $a_i = a_{i-1}+$ a constant. a_i and a_{i-1} can be visualized as two points in some space, a finite distance from each other. In Chapter 6 we studied such potentials. There are other forms of shape invariance such as multiplicative and cyclic, but we shall not consider them here. In the rest of this chapter, when we use the term "shape invariance" we will mean translational shape invariance.

Our fundamental defining equation for shape invariance, Eq. (6.2) is a "difference-differential equation"; that is, it relates the potential, and thus the superpotential W and its spatial derivative computed at two different parameter values a_i and a_{i-1}. Now recall that we have set $\hbar = 1$ in our definition of the potential in terms of the superpotential:

$$V_{\pm}(x,a) = W^2(x,a) \pm \frac{d\,W(x,a)}{dx} \ . \tag{11.1}$$

We have also set the constant relating sequential a's to 1: $a_{i-1} = a_i - 1$. In fact, as we saw when we first defined superpotentials in Chapter 5, the derivative term in the defining equation had an $\hbar$ multiplying it. Until this point we have not needed it, but now we shall. The $\hbar$ will appear in the equation connecting the sequential values of the a's: $a_{i-1} \equiv a_i - \hbar$. In particular, $a_0 \equiv a_1 - \hbar$. But we also saw from all we have done, that inclusion (or exclusion) of $\hbar$ does not affect the physics. *So we do not need to set $\hbar$ to any particular numerical value.* This feature will allow us to take Eq. (6.2) and turn it into a pure differential equation; i.e., one which describes behavior at a single point. This has the obvious advantage of mathematical familiarity (at least to physicists). More important, it provides a new method for finding superpotentials, and thus $g(a_n)$, and from it the energy spectrum of the shape invariant potentials.

Let us simply call a_1 a. Then $a_0 = a - \hbar$, and Eq. (6.2) for shape invariance can now be written

$$W^2(x, a - \hbar) + \hbar \frac{\partial W(x, a - \hbar)}{\partial x} + g(a - \hbar) =$$

$$W^2(x, a) - \hbar \frac{\partial W(x, a)}{\partial x} + g(a). \qquad (11.2)$$

If we now differentiate Eq. (11.2)with respect to $\hbar$, noting that $W^2(x, a)$ and $g(a)$ are independent of $\hbar$, we obtain

$$2 W(x, a - \hbar) \frac{\partial W(x, a - \hbar)}{\partial \hbar} + \frac{\partial W(x, a - \hbar)}{\partial x} +$$

$$\hbar \frac{\partial}{\partial \hbar} \frac{\partial W(x, a - \hbar)}{\partial x} + \frac{\partial g(a - \hbar)}{\partial \hbar} = -\frac{\partial W(x, a)}{\partial x}. \qquad (11.3)$$

Now because of shape invariance, the dependence of W on a and $\hbar$ is through the linear combination $a - \hbar$; therefore, the derivatives of W with respect to a and $\hbar$ are related by

$$\frac{\partial W(x, a - \hbar)}{\partial \hbar} = -\frac{\partial W(x, a - \hbar)}{\partial a}.$$

Note that at this point we have departed from the standard approach of treating x as a continuous variable and a as a discrete parameter. Since the value of $\hbar$ is unspecified, we treat them both on equal footing as continuous variables. Then we may rewrite Eq. (11.3) as

$$-2W(x, a - \hbar) \frac{\partial W(x, a - \hbar)}{\partial a} + \frac{\partial W(x, a - \hbar)}{\partial x} - \hbar \frac{\partial}{\partial a} \frac{\partial W(x, a - \hbar)}{\partial x}$$

$$-\frac{\partial g(a - \hbar)}{\partial a} = -\frac{\partial W(x, a)}{\partial x}.$$

Since this must be true for any value of $\hbar$, we can set $\hbar$ to 0. Then shape invariance implies

$$2 W(x, a) \left(\frac{\partial W(x, a)}{\partial a} \right) - 2 \frac{\partial W(x, a)}{\partial x} + \frac{\partial g(a)}{\partial a} = 0 . \qquad (11.4)$$

This is the desired differential equation which necessarily follows from the original difference-differential equation for shape invariance. Thus, we have shown that all shape invariant superpotentials with $a_0 = a_1 - \hbar$ are solutions of this non-linear partial differential equation.

Before we develop a method to determine solutions of this equation; i.e., to find shape invariant superpotentials, we will give an example of a known shape invariant superpotential, show that it is a solution of this differential equation, and obtain its energy spectrum. Note that there is an important caveat here. Energy eigenvalues are positive in our analysis of shape invariance. This means that $g(a_n) - g(a_0) \geq 0$, whence $\frac{\partial g(A)}{\partial a}$ must be a monotonically increasing function of A.

Scarf I potential: The superpotential is given by

$$W(x, A) = A \tan x + B \sec x \qquad (11.5)$$

where A is the parameter that changes. Substituting this superpotential into Eq. (11.4), we obtain:

$$2 \left(A \tan x + B \sec x \right) (\tan x) - 2 \left(A \sec^2 x + B \sec x \tan x \right) + \frac{\partial g(A)}{\partial A} = 0.$$

I.e., $\frac{\partial g(A)}{\partial A} - 2A = 0$, hence $g(A) = A^2$. To keep this monotonically increasing, we set $a_0 = A$, $a_n = a_0 + n$ and $\hbar = A + n\hbar$. We obtain the spectrum

$$E_n = g(a_n) - g(a_0) = (A + n\hbar)^2 - A^2 .$$

Problem 11.1.

The Morse superpotential is given by

$$W(x, A) = A - Be^{-x}. \qquad (11.6)$$

where A is the parameter that changes additively. Show that this is a solution to the differential equation, and obtain its energy spectrum.

You can check that all of the known shape invariant potentials, listed in the table at the end of this chapter, satisfy the partial differential equation. For a given shape-invariant superpotential, Eq. (11.4) determines $g(a)$, and hence its spectrum.

However, our goal here is the opposite. We will demonstrate a method for generating these superpotentials. Since all shape invariant superpotentials are solutions of Eq. (11.4), solutions of this equation must yield shape invariant superpotentials. As we know from solving the three dimensional Schrödinger equation, the method of separation of variables provided us with all known solutions. Note that the assumption that one can separate variables is just that: an assumption, an ansatz. Here we shall use a variant of this. While we are not necessarily finding all solutions, we hope that a reasonably general ansatz will yield the set of shape invariant potentials. We will thus assume that the superpotential is of the form

$$W(x, a) = \sum_{i=1}^{N} \mathcal{A}_i(a) \, \mathcal{X}_i(x) \qquad (11.7)$$

where $\mathcal{A}_i(a)$ and $\mathcal{X}_i(x)$ are assumed to be real functions of a and x respectively. Thus, to determine $W(x, a)$, we will need to find the functions $\mathcal{A}_i(a)$ and $\mathcal{X}_i(x)$ for various choices of N. We will employ the ansatz for $N = 1$ and $N = 2$.

Throughout, we shall use the following notation:

- Lower case Greek letters will denote "true" (a- and x- independent) constants;
- Upper case Latin letters with a in parentheses will denote a-dependent, but x-independent "constants."

$N=1$: Single-Term Superpotentials

$$W(x, a) = \mathcal{A}(a) \, \mathcal{X}(x) \ .$$

Substituting this ansatz into Eq. (11.4) leads to

$$2 \, (\mathcal{A}(a) \, \mathcal{X}(x)) \left(\mathcal{X}(x) \frac{d \, \mathcal{A}(a)}{da} \right) - 2 \, \mathcal{A}(a) \frac{d \, \mathcal{X}(x)}{dx} + \frac{dg(a)}{da} = 0 \ . \quad (11.8)$$

Combining terms,

$$2 \, \mathcal{A}(a) \left(\mathcal{X}^2 \frac{d \, \mathcal{A}}{da} - \frac{d \, \mathcal{X}}{dx} \right) + \frac{dg}{da} = 0 \ . \quad (11.9)$$

Since $\mathcal{A}$ and $\frac{dg}{da}$ depend only on a, the expression in parentheses, $\left(\mathcal{X}^2 \frac{d \, \mathcal{A}}{da} - \frac{d \, \mathcal{X}}{dx} \right)$ can at most be a function of the parameter a. (It might be a constant independent of a). We set it equal to $H(a)$. For $H(a) \neq 0$ the solution is [1]

$$\mathcal{X} = - \sqrt{\frac{H}{\frac{d \, \mathcal{A}}{da}}} \tanh \left(\sqrt{H \frac{d \, \mathcal{A}}{da}} \, (x - x_0) \right) \ . \quad (11.10)$$

Let us assume for the moment that $H > 0$ Since $\mathcal{X}(x)$ is independent of a, we must have $H(a) \sim \frac{d \, \mathcal{A}(a)}{da}$ and $H(a) \sim \left(\frac{d \, \mathcal{A}(a)}{da} \right)^{-1}$. These two conditions can only be met if both $H(a)$ and $\frac{d \, \mathcal{A}(a)}{da}$ are constants, independent of the parameter a. Thus $\mathcal{A}$ must be a linear function of a: $\mathcal{A} = \mu a + \beta$, and $H(a) = \gamma$ where the Greek-letter constants are all a-independent.

This leads to the superpotential

$$W(x, a) = - (\mu a + \beta) \sqrt{\frac{\gamma}{\mu}} \tanh \left(\sqrt{\gamma \mu} \, (x - x_0) \right) \ .$$

---

[1] For $H(a) = 0$, the solution of the above differential equation is $\mathcal{X} \sim 1/x$ which is a special case of the Coulomb potential with angular momentum $l = 0$. Since we will consider the general case for any l later, we do not discuss it here.

By a scaling ($\gamma\mu = 1$) and a shift ($x_0 = 0$) of x, we can write

$$W(x, a) = A \tanh x , \qquad (11.11)$$

where [2] $A = \left(\frac{\beta}{\mu} - a\right)$. This superpotential is a special case of the Scarf II and the Rosen-Morse superpotentials.

Problem 11.2. *Show that for negative values of H we obtain the superpotential* $W(x, a) = A \tan x$.

For the special case of $\mathcal{A} =$ a pure constant, $H(a)$ must also be a pure constant. This generates

$$W(x, a) = \frac{\omega}{2} x , \qquad (11.12)$$

the superpotential for the *harmonic oscillator*, where we have set $H(a) = -\frac{1}{2}\omega$.

Problem 11.3. *Show this.*

This exhausts the solutions for the $N = 1$ ansatz. Note that by definition it can only give one of the terms in any superpotential. Thus the only complete shape invariant superpotential we find is the harmonic oscillator. All others have two terms [3]. Therefore, we now set $N = 2$. This in fact will generate the rest of the family of known shape invariant superpotentials. The procedure is straightforward but requires a great deal of calculation. We shall just consider a few examples here.

$$W(x, a) = \sum_{i=1}^{2} \mathcal{A}_i(a) \mathcal{X}_i(x) = \mathcal{A}_1(a) \mathcal{X}_1(x) + \mathcal{A}_2(a) \mathcal{X}_2(x) .$$

We would like to very strongly emphasize that neither the functions $\mathcal{A}_1(a)$ and $\mathcal{A}_2(a)$, nor $\mathcal{X}_1(x)$ and $\mathcal{X}_2(x)$ should be linearly proportional to each other, otherwise this case will reduce to the case of $N = 1$ considered above.

Problem 11.4. *Show this.*

Substituting the $N = 2$ ansatz in Eq. (11.4) leads to

$$2\left(\mathcal{A}_1 \mathcal{X}_1 + \mathcal{A}_2 \mathcal{X}_2\right)\left(\mathcal{X}_1 \frac{d\mathcal{A}_1}{da} + \mathcal{X}_2 \frac{d\mathcal{A}_2}{da}\right) + \frac{dg(a)}{da} = \left(\mathcal{A}_1 \frac{d\mathcal{X}_1}{dx} + \mathcal{A}_2 \frac{d\mathcal{X}_2}{dx}\right) .$$

---

[2] Note that this coefficient A is not the same as the function A(a).

[3] Actually, the harmonic oscillator is a special case of the two-term shifted harmonic oscillator: $W(x, a) = \frac{\omega}{2} x - b$.

Expanding it, we obtain:

$$\underbrace{2\mathcal{A}_1 \frac{d\mathcal{A}_1}{da}\,\mathcal{X}_1{}^2}_{\text{Term\#1}} - \underbrace{2\mathcal{A}_1 \frac{d\mathcal{X}_1}{dx}}_{\text{Term\#2}} + \underbrace{2\mathcal{A}_2 \frac{d\mathcal{A}_2}{da}\,\mathcal{X}_2{}^2}_{\text{Term\#3}} - \underbrace{2\mathcal{A}_2 \frac{d\mathcal{X}_2}{dx}}_{\text{Term\#4}} \tag{11.13}$$

$$+ \underbrace{2\,\mathcal{X}_1 \mathcal{X}_2 \left(\mathcal{A}_1 \frac{d\mathcal{A}_2}{da} + \mathcal{A}_2 \frac{d\mathcal{A}_1}{da} \right)}_{\text{Term\#5}} + \underbrace{\frac{dg(a)}{da}}_{\text{Term\#6}} = 0 \ .$$

This is of the form

$$\sum_i^6 F_i(a) G_i(x) = 0 \ ,$$

where

$$
\begin{array}{ll}
F_1 \equiv 2\,\mathcal{A}_1 \, d\mathcal{A}_1/da & G_1 \equiv \mathcal{X}_1^2 \\
F_2 \equiv -2\mathcal{A}_1 & G_2 \equiv d\mathcal{X}_1/dx \\
F_3 \equiv 2\,\mathcal{A}_2 \, d\mathcal{A}_2/da & G_3 \equiv \mathcal{X}_2^2 \\
F_4 \equiv -2\mathcal{A}_2 & G_4 \equiv d\mathcal{X}_2/dx \\
F_5 \equiv 2\,(\mathcal{A}_1 d\mathcal{A}_2/da + \mathcal{A}_2 d\mathcal{A}_1/da) & G_5 \equiv \mathcal{X}_1 \mathcal{X}_2 \\
F_6 \equiv dg/da & G_6 \equiv 1 \ .
\end{array}
\tag{11.14}
$$

This linear combination of various functions of x and a must yield us functions $\mathcal{X}_1$ and $\mathcal{X}_2$ that are independent of a. This severely constrains the $\mathcal{A}_i$'s and $\mathcal{X}_i$'s. These constraints take the form of proportionalities between various terms in Eq. (11.13). The sum of the first five terms must equal $-dg/da$; i.e., that sum must depend at most on a.

The possibilities are:

- Each of the terms 1 through 5 depends on x but their sum does not; i.e., the x dependence disappears. There is only one such possibility.
- One of the terms by itself is independent of x. Then the sum of the other four then also be independent of x. There are five such possibilities, depending on which is the x-independent term.

And so on. As is evident, the fewer the number of terms whose sum is x-independent, the more possibilities there are. But in fact, symmetries eliminate many of the possibilities. Of the five possibilities we just mentioned, due to symmetry between the first and third terms and between the second and fourth, two of the terms are similar to others. For example, the condition "$\mathcal{X}_1$ =constant" is equivalent to "$\mathcal{X}_2$ = constant." [4] This eliminates two possibilities, leaving only three unique ones to examine.

---

[4] As we did in the $N = 1$ case, we must distinguish between constants that depend on a, and those that are a- and x-independent constants. So "constant" here may mean a-dependent or truly constant.

We can characterize the possibilities by separating the terms into subsets which we shall call "irreducibly constant." By this we mean that within a particular subset, say, of three terms whose sum depends only on a, no one term is itself only a-dependent. If it were, then the irreducibly constant subset would have that term removed, and thus consist of the sum of the remaining two.

Even with the constraints and symmetries, there are many cases to examine[5]. Many of the cases produce the same superpotential or reduce to the $N = 1$ case.

Once we have used the symmetries to reduce our possibilities to all sets of irreducibly constant terms, we make the following claim: *If a combination $\sum_i F_i G_i$ is irreducibly constant; i.e., no smaller combination is independent of x, then all F_i's within that combination must be proportional to each other.* Here is the proof.

As we noted earlier, Eq. (11.13) with the definitions of Eq. (11.14) is of the form

$$\sum_i^6 F_i(a)G_i(x) = 0.$$

This is a statement of linear dependence of the coefficients $F_i(a)$ of the vectors $F_i(a)G_i(x)$ in some "G space." Since the sum of all six terms is zero, any one of the terms must be proportional to the sum of the others. So at most five can be linearly independent. Thus, the dimensionality of the space spanned by them would be at most five.

Let us first consider an expression consisting of just two elements: $\{F_1 G_1, F_2 G_2\}$, that add up to a constant irreducibly; i.e., $F_1(a)G_1(x) + F_2(a)G_2(x) = H(a)$. The irreducibility for this case implies that $F_1(a)G_1(x)$ and $F_2(a)G_2(x)$ cannot be separately constant. This implies that neither $G_1(x)$ nor $G_2(x)$ can be x-independent constants, and $F_1(a)$ and $F_2(a)$ cannot be equal to zero, otherwise the two terms would be reducible to one.

Dividing the above equality by $H(a)$ and defining $\mathcal{F}_1(a) = \frac{F_1(a)}{H(a)}$, $\mathcal{F}_2(a) = \frac{F_2(a)}{H(a)}$,

$$\mathcal{F}_1(a)G_1(x) + \mathcal{F}_2(a)G_2(x) = 1. \tag{11.15}$$

Since x and a are real variables, the above expression must be valid for infinitely many values of both x and a. Let us consider a_1 and a_2, two

---

[5]These are all worked out in A. Gangopadhyaya and J.V. Mallow, "Generating shape invariant potentials", *Int. Jour. of Mod. Phys. A* **23**, 4959 (2008).

arbitrarily chosen values of a. For them, we have

$$\mathcal{F}_1(a_1)G_1(x) + \mathcal{F}_2(a_1)G_2(x) = 1$$
$$\mathcal{F}_1(a_2)G_1(x) + \mathcal{F}_2(a_2)G_2(x) = 1. \qquad (11.16)$$

This is a pair of simultaneous equations, with G's the variables and $\mathcal{F}$'s the coefficients. If these can be solved, then each G will be given in terms of $\mathcal{F}$'s, and therefore be x-independent, contrary to our hypothesis. the only way for this not to happen is for

$$\frac{\mathcal{F}_2(a_1)}{\mathcal{F}_1(a_1)} = \frac{\mathcal{F}_2(a_2)}{\mathcal{F}_1(a_2)} . \qquad (11.17)$$

Problem 11.5. *Show this.*

In other words, the ratio $\frac{\mathcal{F}_2(a)}{\mathcal{F}_1(a)}$ is independent of the argument a, and hence must be equal to an a-independent constant. Thus all the F_i's in this two-term irreducibly constant group must be proportional to each other. We can extend the proof to higher dimensional spaces for any of the three-, four-, or five-term irreducible expressions we are considering. The use of these proportionalities is what will lead to the generation of all of the known shape invariant superpotentials.

Let us work out one example, the case where in Eq. (11.13), the sum of the first two terms is irreducibly constant, and each of the remaining terms is itself constant. We can denote this as $\{1, 2\} + \{3\} + \{4\} + \{5\}$.

From $\{1,2\}$, we have $\mathcal{A}_1 \frac{d\mathcal{A}_1}{da} \mathcal{X}_1^2 - \mathcal{A}_1 \frac{d\mathcal{X}_1}{dx} = M(a)$.

Now one possibility is that $\{5\} = 0$ and $M(a) = 0$. From $\{5\} = 0$, since neither $\mathcal{X}_1$ nor $\mathcal{X}_2$ is zero by hypothesis, $\mathcal{A}_1 \frac{d\mathcal{A}_2}{da} + \mathcal{A}_2 \frac{d\mathcal{A}_1}{da} = 0$, whose solution is $\mathcal{A}_1 \sim 1/\mathcal{A}_2$. From $\{3\}$, either $\mathcal{X}_2$ is a constant or $\mathcal{A}_2 \frac{d\mathcal{A}_2}{da} = 0$. Let us take the case $\mathcal{X}_2$ is constant; call it $H(a)$. Since $\{1,2\} = M(a) = 0$, $\mathcal{A}_1 \frac{d\mathcal{A}_1}{da} \mathcal{X}_1^2 = \mathcal{A}_1 \frac{d\mathcal{X}_1}{dx}$.

Integrating over x, $\mathcal{X}_1 = \frac{1}{B - Cx}$ where B is a constant of integration and $C \equiv -\frac{d\mathcal{A}_1}{da}$.

Then the superpotential is $W = \mathcal{A}_1 \mathcal{X}_1 + \mathcal{A}_2 \mathcal{X}_2 = \frac{\mathcal{A}_1}{B - Cx} + H$. B, C, and H are (at most) a-dependent constants. H combines the dependence $\mathcal{A}_1 \sim 1/\mathcal{A}_2$ and the constant $\mathcal{X}_2$. If we set $B = 0$, identify x with radial coordinate r, $\frac{\mathcal{A}_1}{C}$ with $(l + 1)$, and H with $\frac{e^2}{2(l+1)}$, we have the Coulomb superpotential:

$$\frac{e^2}{2(l + 1)} - \frac{(l + 1)}{r}.$$

Problem 11.6. *From* $\{5\} \neq 0$ *and the same irreducibly constant set* $\{1,\ 2\} + \{3\} + \{4\} + \{5\}$, *obtain the superpotential for the three-dimensional harmonic oscillator,* $W = \frac{\omega r}{2} - \frac{l+1}{r}$.

The table lists all of the known one- and two-term shape invariant superpotentials and their potentials. Next to each is the value of N. The only case which is solved by N=1 is the harmonic oscillator, the only superpotential which has one term (ignoring the trivial shift $-b$ mentioned in the earlier footnote). All others are obtained from $N = 2$. $N = 1$ will give one but not both of the terms of these superpotentials We saw this as we obtained from $N = 1$ the superpotential $W(x, a) = A \tanh x$, the first term of the Scarf II and the Rosen-Morse superpotentials. The final column lists one combination of irreducibly constant groups from which the superpotential was obtained. These combinations are not unique. There are numerous combinations which will generate the same superpotential. In fact, there are many which reduce to the harmonic oscillator. Each combination must be tested. We can see that even for $N = 2$ there is a great deal of work involved. If we try $N = 3$ the labor becomes prohibitive: there are more than a thousand possible combinations. However, the method is a paradigm for how to generate superpotentials, and some variant might be applied to obtain superpotentials which are multiplicatively or cyclically invariant.

Name	Superpotential	N	Combination
Harmonic Oscillator	$\frac{1}{2}\omega x$	1	
Coulomb	$\frac{e^2}{2(l+1)} - \frac{l+1}{r}$	2	$\{1,2\}+\{3\}+\{4\}+(\{5\}=0)$
3D-harmonic oscillator	$\frac{1}{2}\omega r - \frac{l+1}{r}$	2	$\{1,2\}+\{3\}+\{4\}+(\{5\}\neq 0)$
Morse	$A - Be^{-x}$	2	$\{2,5\}+\{1\}+\{3\}+\{4\}$
Pöschl-Teller I	$A\tan r - B\cot r$	2	$\{1,2\}+\{3,4\}+(\{5\}\neq 0)$
Pöschl-Teller II	$A\tanh r - B\coth r$	2	$\{1,2\}+\{3,4\}+(\{5\}\neq 0)$
Rosen-Morse I	$-A\cot x - \frac{B}{A}$	2	$\{1,2\}+\{3\}+\{4\}+(\{5\}=0)$
Rosen-Morse II	$A\tanh x + \frac{B}{A}$	2	$\{1,2\}+\{3\}+\{4\}+(\{5\}=0)$
Eckart	$-A\coth x + \frac{B}{A}$	2	$\{1,2\}+\{3\}+\{4\}+(\{5\}=0)$
Scarf I	$A\tan x - B\sec x$	2	$\{1,2\}+\{4,5\}+\{3\}$
Scarf II	$A\tanh x + B\operatorname{sech} x$	2	$\{1,2\}+\{4,5\}+\{3\}$
Generalized Pöschl-Teller	$A\coth x - B\operatorname{cosech} x$	2	$\{1,2\}+\{4,5\}+\{3\}$

Chapter 12

Singular Potentials in Supersymmetric Quantum Mechanics

In this chapter we will study systems described by singular potentials, as they are abundant in quantum mechanics. For these systems, the value of the potential becomes unbounded either at a set of points known as the singular points or over a region. In some sense, the term "singular *potential*" is a misnomer: the actual issue is not *whether* the potential blows up, but *how* it does so. As we shall show shortly, only potentials with a singular term of the form α/r^2 need concern us. Such terms often originate from the non-zero angular momentum of a system. It is the particular value of the coefficient α that allow these systems to be solved. Similar "fortuitous" coefficients appear in other potentials as well. In addition to the non-zero angular momentum type, potentials such as the Rosen-Morse, Eckart, and Pöschl-Teller also have this kind of singularity. We will find that in these cases $\alpha \geq -\frac{1}{4}$. Otherwise, the particle would be "sucked into" the singularity; i.e., in such cases there is no lower bound on the ground state energy as we approach the singularity. Compare this for example to the hydrogen atom, where despite the potential becoming unboundedly negative as $r \to 0$, the ground state energy is a well-defined $-13.6\ eV$. We will see that supersymmetric quantum mechanics provides natural barriers so that potentials with $\alpha < -\frac{1}{4}$ simply do not arise.

In a one dimensional system, a term of the type α/x^2 has the potential of breaking the entire real line into two non-communicating halves. This, as we shall see, happens if $\alpha \geq \frac{3}{4}$. In such a case a particle situated on one side of the singularity never tunnels to the other side. We will also see that for $\alpha \geq \frac{3}{4}$, the eigenvalue problem is well defined and can be solved by conventional means. For these one- and three dimensional cases we say that we have a "hard" singularity.

Problem 12.1. *Determine the value of α for these three superpotentials.*

$$\text{Rosen} - \text{Morse} : \ W(x, A, B) \ = \ - A \cot x - B/A$$
$$\text{Eckart} : \ W(x, A, B) \ = \ - A \coth r + B/A$$
$$\text{Pöschl-Teller:} \ W(x, A, B) \ = \ - A \coth r - B \operatorname{cosech} r$$

For $-\frac{1}{4} \leq \alpha < \frac{3}{4}$, the behavior of both solutions at the origin is such that they are square integrable. Hence, the boundary conditions are not sufficient to determine eigenvalues and eigenfunctions, and there is no *a priori* mechanism to select any specific linear combination. Thus, in the context of ordinary quantum mechanics, the spectrum of the system can be determined only by imposing additional ad hoc constraints [1]. These are known as "soft" singularities. They neither suck the particle in, nor do they break up the real axis. Potentials with this type of soft singularity are called "transition potentials." We will show that in some cases these transition potentials emerge naturally in supersymmetric quantum mechanics.

We thus have two problems to address: Why is the inverse square potential the only one we need consider, and why for this case alone are there constraints on α? The answer to both questions was provided in a semi-quantitative way (Landau and Lifshitz cited in footnote 1). The mean energy of a system is the sum of its average kinetic and potential energies: $E_{\text{mean}} = \langle T \rangle + \langle V \rangle$. Let us assume that the wave function is appreciable mostly within a distance r_0 of the origin. Then from uncertainty arguments we can make the following estimates: since the particle's position $\Delta x \sim r_0$, its momentum uncertainty is of order $\Delta p \sim \hbar/r_0$. Treating Δp as a standard deviation, $\Delta p^2 = \langle p^2 \rangle - \langle p \rangle^2$. But $\langle p \rangle = 0$ because it takes into account the opposite directions of p. Thus, the average kinetic energy $\langle T \rangle = \langle p^2 \rangle/2m \simeq \hbar^2/2mr_0{}^2$. Let us assume that the potential obeys an inverse power law: $V = -\alpha/r^s$. Now let us consider what happens for various values of s. For $s > 2$, V is the dominant term as $r_0 \rightarrow 0$. The mean energy is therefore unbounded, so there can be no finite ground state. This is the case of the particle being sucked in to the singularity. Conversely, for $s < 2$ the energy is bounded at some finite negative value. Again, we mention as an example the hydrogen atom, with well-defined ground state energy. It is only for the case $s = 2$ that the kinetic and potential energy

---

[1]See L. D. Landau and E. M. Lifshitz, *Quantum Mechanics*, NY: Pergamon Press (1977); W. M. Frank, D. J. Land and R. M. Spector, "Singular potentials", *Rev. Mod. Phys.* **43**, 36 (1971); and A. Gangopadhyaya, P. K. Panigrahi and U. P. Sukhatme, "Analysis of inverse-square potentials using supersymmetric quantum mechanics", *J. Phys.* **A27** 4295 (1994) .

terms are of the same order near r_0. Therefore the spectrum will depend on the magnitude of α as compared with $\hbar^2/2m$.

To summarize, we need only consider the case $s = 2$. For that case, we will show the following: the particle will be sucked into the singularity when $\alpha < -\frac{1}{4}$; i.e., there is no finite-energy ground state. For $\alpha > \frac{3}{4}$, the eigenvalues and eigenfunctions can be obtained from the normalization condition. However, for $-\frac{1}{4} \leq \alpha \leq \frac{3}{4}$, the problem cannot be solved by conventional quantum mechanics, but only with the imposition of additional constraints. SUSYQM on the other hand will provide the eigenvalues and eigenfunctions without the need for additional constraints needed in ordinary quantum mechanics. The various situations are illustrated in Figure 12.1.

Let us study the above statements through an example. We consider a system described by a potential with an α/r^2 singularity. The radial part $\mathcal{R}(r)$ of the three-dimensional wave function obeys

$$-\frac{1}{r^2}\frac{d\mathcal{R}}{dr}\left(r^2\frac{d\mathcal{R}}{dr}\right) + V(r)\mathcal{R} = E\mathcal{R} \ . \tag{12.1}$$

We define a new function [2] $\psi = r\,\mathcal{R}$. The equation for ψ then becomes an effective one-dimensional Schrödinger equation

$$-\frac{d^2\psi}{dr^2} + V(r)\psi = E\psi(r) \ . \tag{12.2}$$

The normalization of this function ψ is derived from the normalization of $\mathcal{R}$:

$$\int_0^\infty |\mathcal{R}|^2\,r^2 dr = \int_0^\infty |\psi|^2_-\,dr = 1 \ .$$

For the last integral to be finite, it is necessary that the behavior of $|\psi|^2$ near the origin be less divergent than the function $\frac{1}{r}$; i.e., ψ should not blow up at the origin any faster than $\frac{1}{\sqrt{r}}$.

Problem 12.2. *Show that normalizability of $\psi(r)$ requires that the function be less singular than $\frac{1}{\sqrt{r}}$ near the origin.*

In the vicinity of the origin, finite terms such as $E\,\psi$ are much smaller than the α/r^2 term and can be ignored. Eq. (12.2) then reduces to

$$-\frac{d^2\psi}{dr^2} + \frac{\alpha}{r^2}\,\psi \approx 0 \ . \tag{12.3}$$

---

[2] We called these u and R in our discussion of the hydrogen in Chapter 2, to stay with customary notation in most quantum mechanics texts.

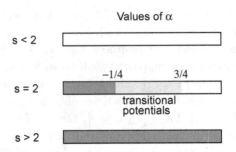

Fig. 12.1 The lightly shaded region depicts the range of α $\left(-\frac{1}{4} \leq \alpha \leq \frac{3}{4}\right)$ that allows for the ambiguity in determination of the spectrum. For $s < 2$ there is no ambiguity, and for the domain $s > 2$ there are no bound states.

Multiplying Eq. (12.3) by r^2 and solving the resulting equation, we find that all eigenfunctions near the origin reduce to the linear combination

$$\psi \approx c_1 \, r^{\left(\frac{1}{2}-\sqrt{\alpha+\frac{1}{4}}\right)} + c_2 \, r^{\left(\frac{1}{2}+\sqrt{\alpha+\frac{1}{4}}\right)} \ . \tag{12.4}$$

Problem 12.3. *Show that ψ near the origin reduces to the combination given in Eq. (12.4).*

For $\alpha < -\frac{1}{4}$, both terms of Eq. (12.4) become oscillatory and thus cannot generate a nodeless ground state.

For $\alpha \geq \frac{3}{4}$, the first term is not normalizable and hence $c_1 = 0$. c_2 is then obtained from the normalization. In addition, the wave function must vanish at at the non-singular end of the domain. This quantizes the energy. Hence, the spectrum is determinable for $\alpha \geq \frac{3}{4}$.

Now, let us explore the transitional region $-\frac{1}{4} \leq \alpha < \frac{3}{4}$. For α within this range, both linearly independent solutions of the Schrödinger equation are less divergent that $\frac{1}{\sqrt{r}}$ near the origin, and hence normalizable.

Problem 12.4. *Show that for $-\frac{1}{4} \leq \alpha < \frac{3}{4}$ both terms of Eq. (12.4) are less singular than $\frac{1}{\sqrt{r}}$.*

Hence we cannot say *a priori* which is the correct eigenfunction. For this we must find a way to impose additional constraints. There are several ways to do so; we shall outline one here. These methods turn out to select the term with the higher power of r as the correct one. Conventionally, this higher power term is called the "less singular" function. For negative powers of r, when at least one term is singular, this is meaningful; the terminology is used even when neither of the terms is singular.

This approach of retaining the less singular wave function then leads to the determination of eigenvalues and eigenfunctions. But let us remark that, as we shall show later, SUSYQM makes this selection naturally; i.e., with no need for extra constraints.

Using the method of the reference in footnote 1, we first "regularize" the potential by taking it to be a constant within a small neighborhood of radius r_0 around the singular point. I.e., for $r < r_0$, the Schrödinger equation is

$$-\frac{d^2\psi}{dr^2} + \frac{\alpha}{r_0^2}\psi = 0 \ , \qquad (12.5)$$

and its solutions are given by

$$\psi = A \sinh\left(\frac{\sqrt{\alpha}}{r_0}\, r\right) + B \cosh\left(\frac{\sqrt{\alpha}}{r_0}\, r\right).$$

Problem 12.5. *Derive the above solution for $r < r_0$.*

Since ψ vanishes at the origin, we must set $B = 0$. Hence the solution inside the regularized region is $A \sinh\left(\frac{\sqrt{\alpha}}{r_0}\, r\right)$. The solution for $r > r_0$, is

$$\psi = c_1\, r^{\left(\frac{1}{2} - \sqrt{\alpha + \frac{1}{4}}\right)} + c_2\, r^{\left(\frac{1}{2} + \sqrt{\alpha + \frac{1}{4}}\right)}.$$

Problem 12.6. *Derive the above solution for $r > r_0$.*

Matching the boundary conditions for the wave function and its derivative and dividing one by the other [3], we obtain

$$\sqrt{\alpha}\coth\left(\sqrt{\alpha}\right) = \frac{\left(\frac{1}{2} - \sqrt{1 + 4\alpha}\right)\frac{c_1}{c_2} + \left(\frac{1}{2} + \sqrt{1 + 4\alpha}\right) r_0^{2\sqrt{1+4\alpha}}}{\frac{c_1}{c_2} + r_0^{2\sqrt{1+4\alpha}}} \ . \qquad (12.6)$$

Solving for $\frac{c_1}{c_2}$ we obtain

$$\frac{c_1}{c_2} = \frac{r_0^{2\sqrt{1+4\alpha}}\left(\frac{1}{2} + \sqrt{1 + 4\alpha} - \sqrt{\alpha}\coth\left(\sqrt{\alpha}\right)\right)}{\left(\sqrt{\alpha}\coth\left(\sqrt{\alpha}\right) - \frac{1}{2} + \sqrt{1 + 4\alpha}\right)} \ . \qquad (12.7)$$

Taking the limit $r_0 \to 0$, we find that $c_1 \to 0$, which means that the less singular wave function gets selected as the proper ground state eigenfunction [4].

We will now show that SUSYQM provides an alternate way of arriving at the same prescription without the need for regularization. To see how

---

[3] This is the logarithmic derivative, familiar from the infinite square well.

[4] For more details, see footnote 1.

this happens let us start with an appropriate superpotential near the origin $W(r, \alpha) \approx -\frac{(1+\sqrt{1+4\alpha})}{2r}$ such that the potential $V_-(r, \alpha) \approx \alpha/r^2$ falls in the transitional region $-\frac{1}{4} \le \alpha < \frac{3}{4}$. We will then show that the eigenspectrum of the supersymmetric partner potential $V_+(r, \alpha)$ can be determined unambiguously.

Solving for V_+ first and then using the SUSYQM relations $E_{n+1}^{(-)} = E_n^{(+)}$, and $\psi_{n+1}^{(-)} = A^+ \psi_n^{(+)}$, we can solve the eigenvalue problem for the potential V_-. Thus, SUSYQM provides a straightforward prescription for choosing the "less singular" solution.

Problem 12.7. *Show that for* $W(r, \alpha) = -\frac{(1+\sqrt{1+4\alpha})}{2r}$, *the partner potentials are* $V_-(r, \alpha) = \frac{\alpha}{r^2}$ *and* $V_+(r, \alpha) = \frac{(1+\sqrt{1+4\alpha})(3+\sqrt{1+4\alpha})}{4r^2}$.

Since near the origin the superpotential $W(r, \alpha)$ is given by $W(r, \alpha) \approx -\frac{(1+\sqrt{1+4\alpha})}{2r}$, the resulting partner potentials are

$$V_-(r, \alpha) \approx \frac{\alpha}{r^2} \quad ; \quad V_+(r, \alpha) \approx \frac{(1 + \sqrt{1+4\alpha})(3 + \sqrt{1+4\alpha})}{4r^2}.$$

Their eigenfunctions are

$$\psi^{(-)}(r, \alpha) \approx c_1\, r^{\left(\frac{1}{2} - \sqrt{\alpha + \frac{1}{4}}\right)} + c_2\, r^{\left(\frac{1}{2} + \sqrt{\alpha + \frac{1}{4}}\right)} \tag{12.8}$$

and

$$\psi^{(+)}(r, \alpha) \approx c_1'\, r^{-\left(\frac{1}{2} + \sqrt{\alpha + \frac{1}{4}}\right)} + c_2'\, r^{\left(\frac{3}{2} + \sqrt{\alpha + \frac{1}{4}}\right)}. \tag{12.9}$$

From Eq. (12.8) it is clear that for $-\frac{1}{4} \le \alpha < \frac{3}{4}$, both linearly independent solutions are less divergent than $\frac{1}{\sqrt{r}}$ near the origin. Hence, $\psi^{(-)}(r, \alpha)$ is square integrable. Thus, normalization does not place any constraints on values of c_1 and c_2.

However, with the value of α within this transitional region, one of the terms in Eq. (12.9) is normalizable. Thus, the requirement of square integrability of $\psi^{(+)}(r, \alpha)$ uniquely determines the proper linear combination near the origin, and hence unambiguously determines the eigenvalues and eigenfunctions for the partner potential V_+. Each eigenfunction $\psi^{(-)}(r, \alpha)$ can then be obtained by applying operator A^+ to the corresponding eigenfunction $\psi^{(+)}(r, \alpha)$ of V_+. Eigenvalues of V_- will be identical to those of V_+ with the exception of an additional zero energy ground state for V_- in case of unbroken SUSY.

We will now work out a full fledged example with the potential given over the entire domain and not just near the singularity.

Let us consider the Pöschl-Teller II superpotential

$$W(r, A, B) = A \tanh r \; - \; B \coth r \; , \quad 0 \le r < \infty \; . \tag{12.10}$$

For $A > B$ and $B > 0$ the above superpotential corresponds to a case of unbroken SUSY as discussed in Chapter 5. The corresponding supersymmetric partner potentials are

$$V_-(r, A, B) = -A(A+1)\operatorname{sech}^2 r \; + \; B(B-1)\operatorname{cosech}^2 r \; + \; (A-B)^2 \; ,$$
$$V_+(r, A, B) = -A(A-1)\operatorname{sech}^2 r \; + \; B(B+1)\operatorname{cosech}^2 r \; + \; (A-B)^2 \; . \tag{12.11}$$

The above set of potentials is shape invariant under the change of parameters $A \to A - 1$ and $B \to B + 1$. I.e., $V_+(r, A, B) = V_+(r, A-1, B+1) +$ constant. Without loss of generality we will assume $\frac{1}{2} < A < \infty$, and $\frac{1}{2} \le B < \infty$. To clearly see the ambiguity in the eigenvalue problem, we proceed with the analysis of the Schrödinger equation for $V_-(r)$. The general solution is

$$\psi^{(-)}(r) = \cosh^{-A} r \left[c_1 \sinh^B r F\left(a', b', c'; -\sinh^2 r\right) \right.$$

$$\left. + c_2 \sinh^{(1-B)} r F\left(a' + 1 - c', b' + 1 - c', 2 - c'; -\sinh^2 r\right) \right], \tag{12.12}$$

where the constants a', b', and c' are given by

$$a' = \frac{1}{2}(B - A + Q') \; ; \; b' = \frac{1}{2}(B - A - Q') \; ; \; c' = B + \frac{1}{2} \; ; \; \text{and}$$

$$Q' = \sqrt{(B-A)^2 - E} \; . \tag{12.13}$$

Near the point $r = 0$, the solution reduces to

$$\psi^{(-)}(r) \approx c_1 \, r^B + c_2 \, r^{(1-B)} \; . \tag{12.14}$$

Thus for $B \ge \frac{3}{2}$ the wave function becomes non-normalizable unless $c_2 = 0$. With $c_2 = 0$, a subsequent constraint: the vanishing of the wave function at infinity as demanded by the normalizability, suffices to determine the eigenvalues E in terms of the parameters A and B. However, if $\frac{1}{2} \le B < \frac{3}{2}$, both terms of Eq. (12.14) are well defined. Hence normalizability places no constraints on their coefficients. In such cases, we solve the eigenvalue problem for V_+ instead. Its eigenfunctions near the origin are of the form:

$$\psi^{(+)}(r) \approx \tilde{c}_1 \, r^{(B+1)} + \tilde{c}_2 \, r^{(-B)} \; . \tag{12.15}$$

Clearly, the normalizability of $\psi^{(+)}$ for $B < \frac{3}{2}$ requires that we set $\tilde{c}_2 = 0$. To determine the eigenvalues of V_+, we have to study the behavior of

$\psi^{(+)}(r)$ at infinity. Using an alternate asymptotic form of the hypergeometric function:

$$F(a, b, c; z) \approx \frac{\Gamma(c)\Gamma(b-a)}{\Gamma(b)\Gamma(c-a)} (-z)^{-a} + \frac{\Gamma(c)\Gamma(a-b)}{\Gamma(a)\Gamma(c-b)} (-z)^{-b} , \qquad (12.16)$$

we find

$$\psi^{(+)}(r) \approx \frac{\Gamma(c)\Gamma(b-a)}{\Gamma(b)\Gamma(c-a)} e^{-Qr} + \frac{\Gamma(c)\Gamma(a-b)}{\Gamma(a)\Gamma(c-b)} e^{Qr} \qquad (12.17)$$

where the constants a, b, c and Q are obtained from Eq. (12.13) by replacing A by $A-1$ and B by $B+1$. They are given by

$$a = \frac{1}{2}(B - A + 2 + Q) \ ; \ b = \frac{1}{2}(B - A + 2 - Q) \ ; \ c = B + \frac{3}{2} \ ;$$

$$Q = \sqrt{(B - A + 2)^2 - \bar{E}} \ ; \ \bar{E} = E + (A - B - 2)^2 - (A - B)^2 . \qquad (12.18)$$

The second term on the right hand side of Eq. (12.17) must vanish to have a well defined bound state. This can be achieved if either a or $(c - b)$ is equal to a negative number; say $-k$. [5] If $a = -k$, then the eigenvalues of V_+ are given by [6]

$$E_k^{(+)} = (A - B)^2 - [A - B - 2k - 2]^2 \quad k = 0, 1, \ldots, n, \qquad (12.19)$$

where n is the number of bound states that the potential will hold. It is the largest integer such that $Q > 0$ i.e. $A - B - 2 > 2n$.

The eigenvalues for V_- will be the same as for V_+, except that V_- will have an additional state (its ground state) with zero energy. The eigenfunctions of V_+ are given by

$$\psi^{(+)}(r) = (\sinh r)^{1+B} (\cosh r)^{A+1} \times$$
$$F\left(-k, \frac{1}{2}(B - A + 2 - Q), B + \frac{3}{2} ; -\sinh^2 r\right). \qquad (12.20)$$

It is customary to express these eigenfunctions in terms of the Jacobi polynomials P_k, which are related to hypergeometric functions F. In terms of Jacobi polynomials, the eigenfunctions are given by

$$\psi_k^{(+)}(r) = (\sinh r)^{(1+B)} (\cosh r)^{-(A-1)} P_k^{(B+\frac{1}{2}, -A+\frac{1}{2})} (\cosh 2r) ,$$
$$k = 0, 1, \ldots, n . \qquad (12.21)$$

---

[5] The inverse of the Gamma function Γ is zero at negative integers.

[6] If instead, the second condition holds; i.e., $c - b = -k$, the eigenvalues are given by $E_k^{(-)} = (A - B)^2 - [A + B + 2k + 1]^2$, $k = 0, 1, \ldots, n$. The condition on A for n-bound states in the second case, obtained by requiring that $Q > 0$, is given by $A < -B - 2n - 1$, which cannot be satisfied for any n, as we have assumed $-\frac{1}{2} \le A < \infty$ and $\frac{1}{2} \le B < \infty$.

The eigenfunctions of the system with potential V_- are now obtained by applying the operator A^+ to the function $\psi^{(+)}$. Near the origin, $\psi^{(+)}(r) \approx r^{B+1}$. Operating with A^+ on $\psi^{(+)}$ lowers the power of r by one, and hence $\psi^{(-)}(r) \approx r^B$. Comparing this result with Eq. (12.14), we see that SUSYQM automatically chooses the term with higher power of r. This is also the "less singular" of the two terms, and is consistent with the prescription used in references given in footnote 1. Thus, SUSYQM naturally selects the less divergent term without any need for additional assumption such as regularization. We find that the SUSYQM based formalism gives a clear-cut way of finding the eigenvalues and eigenfunctions of transition potentials in the region of ambiguity.

Now we will show that there are cases where α/x^2-type interactions appear naturally. There is an even more interesting fact about these potentials: They are necessarily of transitional nature; i.e., the value of α automatically fall within the region $-\frac{1}{4} \leq \alpha < \frac{3}{4}$.

As we have seen in Chapter 6, in any shape invariant system the eigenvalues can be generated by

$$E_0^{(-)} = 0 \ , \ E_n^{(-)} = \sum_{k=0}^{n-1} R(a_k) \ , \tag{12.22}$$

and their corresponding eigenfunctions are given by

$$\psi_0^{(-)} \propto e^{-\int_{x_0}^x W(y,a_0)dy} \ , \ \psi_n^{(-)}(x,a_0) = \left[-\frac{d}{dx} + W(x,a_0) \right] \psi_{n-1}^{(-)}(x,a_1) \ ,$$
$$n = 1,2,3,\ldots \tag{12.23}$$

where $a_1 = f(a_0), \ldots a_k = f^k(a_0)$. If f is the identity function; i.e., $f^2(a_0) = a_0$, we obtain the equally spaced spectrum of the harmonic oscillator. We now consider a function such that $f(a_0) = a_1 \neq a_0$, but $f(a_1) = a_0$. We can derive a potential that is generated by such a function and determine its eigenvalues and eigenfunctions. Since $f^2(a_0) = a_0$, we have

$$a_0 = a_2 = a_4 = \ldots \ , \text{ and } \ f(a_0) = a_1 = a_3 = a_5 = \ldots \ .$$

There are many functions which meet the above requirement. One such function is $f(a_0) = C/a_0$, where C is an arbitrary constant. Another simple possibility is $f(a_0) = -a_0$. A less trivial transformation is $f(a_0) = \frac{1}{2C} \log[\coth(Ca_0)]$. From Eq. (12.22), the eigenvalues will be spaced alternately by $R(a_0) \equiv \omega_0$ and $R(a_1) \equiv \omega_1$. This spectrum corresponds to two shifted sets of equally spaced eigenvalues.

To explicitly compute the superpotential associated with these eigenvalues, we need to solve the shape invariance equations

$$W^2(x, a_0) + W'(x, a_0) = W^2(x, a_1) - W'(x, a_1) + \omega_0 \ , \tag{12.24}$$

$$W^2(x, a_1) + W'(x, a_1) = W^2(x, a_0) - W'(x, a_0) + \omega_1 \ . \tag{12.25}$$

After some algebra we find that the superpotential $W(x, a_0)$ is given by

$$W(x, a_0) = \frac{1}{2}\,\omega\,x + \frac{1}{2}\frac{\Omega}{\omega}\frac{1}{x} \ , \tag{12.26}$$

where $\omega = \frac{1}{2}(\omega_0 + \omega_1)$ and $\Omega = \frac{1}{2}(\omega_0 - \omega_1)$. This gives $\omega_0 = (\Omega + \omega)$ and $\omega_1 = (\omega - \Omega)$.

Problem 12.8. *Derive the above superpotential.*

The corresponding potential $V_-(x)$ is

$$V_-(x, a_0) = \frac{1}{4}\frac{\Omega(\Omega + 2\omega)}{\omega^2}\frac{1}{x^2} + \frac{1}{4}\,\omega^2 x^2 - \frac{(\omega - \Omega)}{2}. \tag{12.27}$$

Problem 12.9. *Show that for positive values of ω_0 and ω_1, $\frac{1}{4}\frac{\Omega(\Omega + 2\omega)}{\omega^2}$ is always between $-\frac{1}{4}$ and $\frac{3}{4}$.*

Since for all positive values of ω_0 and ω_1, the coefficient of the $1/x^2$-term is constrained by $-\frac{1}{4} \leq \left(\frac{1}{4}\frac{\Omega(\Omega + 2\omega)}{\omega^2}\right) < \frac{3}{4}$, the potential of Eq. (12.27) is necessarily of the transitional type; i.e., a particle on the left can tunnel through the singularity and vice versa.

However, there is a problem. The superpotential $W(x, a_0)$ has an infinite discontinuity at the origin. Such a discontinuity is not acceptable, since while writing Eqs. (12.24) and (12.25), we implicitly assumed the existence of a continuous superpotential with a well-defined derivative. To remedy this we regularize the superpotential. We construct a continuous superpotential $\widetilde{W}(x, a_0, \epsilon)$ which reduces to $W(x, a_0)$ in the limit $\epsilon \to 0$. One such choice is

$$\widetilde{W}(x, a_0, \epsilon) = W(x, a_0)\,\tanh^2 \frac{x}{\epsilon} \ . \tag{12.28}$$

The moderating factor $\tanh^2 \frac{x}{\epsilon}$ provides a smooth interpolation through the discontinuity, since it is unity everywhere except in a small region of order ϵ around $x = 0$. In this region, $\widetilde{W}(x, a_0, \epsilon)$ is linear with a slope $\frac{1}{2\epsilon^2}\frac{\omega}{\Omega}$, which becomes larger as ϵ gets smaller. The potential $\widetilde{V}_-(x, a_0, \epsilon)$ corresponding to the superpotential $\widetilde{W}(x, a_0, \epsilon)$ is

$$\widetilde{V}_-(x, a_0, \epsilon) = \widetilde{W}^2(x, a_0, \epsilon) - \widetilde{W}'(x, a_0, \epsilon) \ . \tag{12.29}$$

In the limit $\epsilon \to 0$, $\frac{1}{2\epsilon} \operatorname{sech}^2 \frac{x}{\epsilon} = \delta(x)$ and $\tanh \frac{x}{\epsilon} = [2\theta(x) - 1]$ where $\theta(x)$ is the unit step (Heaviside) function. Thus, $\widetilde{V}_-(x, a_0, \epsilon)$ reduces to

$$\widetilde{V}_-(x, a_0) = V_-(x, a_0) - 4\, W(x, a_0)\, [2\theta(x) - 1]\, \delta(x) \; . \qquad (12.30)$$

Note that the potential $\widetilde{V}_-(x, a_0)$ has an an additional singularity at the origin given by $-\frac{2}{\omega}\frac{\Omega}{x} \frac{1}{x} [2\theta(x) - 1]\, \delta(x)$. The sign of this singularity depends upon the values of ω_0 and ω_1. Let us consider the two cases $\omega_0 > \omega_1$ and $\omega_0 < \omega_1$ separately.

Case 1: $\omega_0 > \omega_1$. An example of a superpotential for this case is shown in Figure 12.2. The corresponding potential is drawn in Figure 12.3, along with a few low lying energy levels. Note the sharp attractive shape of the potential near $x = 0$. In the limit $\epsilon \to 0$, this attractive δ-function singularity produces a bound state at $E_0 = 0$.

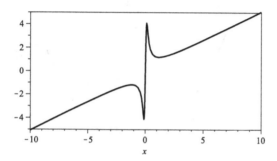

Fig. 12.2   The superpotential $\widetilde{W}(x, a_0, \epsilon)$ given by Eq. (12.28) for $\epsilon = 0.1$ and parameters $\omega_0 = 1.7, \omega_1 = 0.3$.

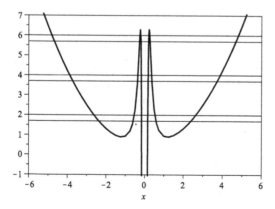

Fig. 12.3   The potential $\widetilde{V}_-(x, a_0)$ given by Eq. (12.30) for $\epsilon = 0.1$ and parameters $\omega_0 = 1.7, \omega_1 = 0.3$.

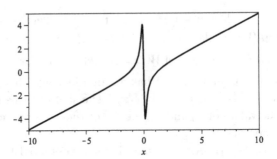

Fig. 12.4 The superpotential $\widetilde{W}(x, a_0, \epsilon)$ given by Eq. (12.28) for $\epsilon = 0.1$ and parameters $\omega_0 = 0.3, \omega_1 = 1.7$.

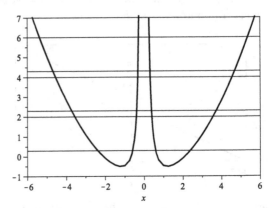

Fig. 12.5 The potential $\widetilde{V}_-(x, a_0)$ given by Eq. (12.30) for $\epsilon = 0.1$ and parameters $\omega_0 = 0.3, \omega_1 = 1.7$.

Case 2: $\omega_0 < \omega_1$. An example of a superpotential for this case is shown in Figure 12.4. Again, proceeding as in case 1, we find a repulsive singularity in the potential at the origin. This is shown in Figure 12.5.

We obtain eigenvalues and eigenfunctions for the potential $\widetilde{V}_-(x, a_0)$ from Eq. (12.23). The lowest four are:

$$E_0 = 0; \qquad \psi_0 \propto x^{-\frac{\omega_0 - \omega_1}{2(\omega_0 + \omega_1)}} \, e^{-\frac{1}{8}(\omega_0 + \omega_1)x^2} \, ,$$

$$E_1 = \omega_0; \qquad \psi_1 \propto x^{1 + \frac{\omega_0 - \omega_1}{2(\omega_0 + \omega_1)}} \, e^{-\frac{1}{8}(\omega_0 + \omega_1)x^2} \, ,$$

$$E_2 = \omega_0 + \omega_1;$$

$$\psi_2 \propto \left(\frac{\omega_0 - \omega_1}{\omega_0 + \omega_1} - 1 + \frac{\omega_0 + \omega_1}{2}x^2 \right) x^{-\frac{\omega_0 - \omega_1}{2(\omega_0 + \omega_1)}} \, e^{-\frac{1}{8}(\omega_0 + \omega_1)x^2} \, ,$$

$$E_3 = 2\omega_0 + \omega_1;$$

$$\psi_3 \propto \left(-\frac{\omega_0 - \omega_1}{\omega_0 + \omega_1} - 3 + \frac{\omega_0 + \omega_1}{2}x^2 \right) x^{1 + \frac{\omega_0 - \omega_1}{2(\omega_0 + \omega_1)}} \, e^{-\frac{1}{8}(\omega_0 + \omega_1)x^2} \, .$$

In Figures 12.6 and 12.7, we have drawn the lowest few eigenfunctions for two choices of the parameters ω_0 and ω_1. Note that for $\omega_0 > \omega_1$, the ground state eigenfunction diverges at the origin, whereas for $\omega_0 < \omega_1$ it vanishes. For the intermediate situation $\omega_0 = \omega_1$, the ground state is finite at the origin and is just the standard Gaussian solution of a one dimensional oscillator.

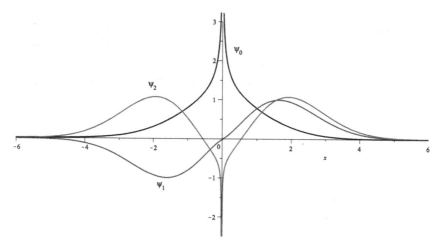

Fig. 12.6   The first three eigenfunctions corresponding to the potential of Eq. (12.30) for $\omega_0 = 1.7, \omega_1 = 0.3$.

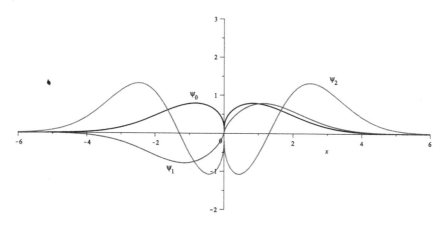

Fig. 12.7   The first three eigenfunctions corresponding to the potential of Eq. (12.30) for $\omega_0 = 0.3, \omega_1 = 1.7$.

General expressions for these eigenfunctions and corresponding eigenergies are given by

$$E_{2n} = n(\omega_0 + \omega_1) , \qquad \psi_{2n} \propto x^{-\frac{d}{2}} e^{-\frac{1}{4}\omega x^2} L_n^{-\frac{d}{2}-\frac{1}{2}} \left[\frac{\omega x^2}{2}\right] ,$$
$$E_{2n+1} = n(\omega_0 + \omega_1) + \omega_0 , \quad \psi_{2n+1} \propto x^{1+\frac{d}{2}} e^{-\frac{1}{4}\omega x^2} L_n^{\frac{d}{2}+\frac{1}{2}} \left[\frac{\omega x^2}{2}\right] , \qquad (12.32)$$

where $d \equiv \frac{\omega_0 - \omega_1}{(\omega_0 + \omega_1)}$, $\omega \equiv \frac{(\omega_0 + \omega_1)}{2}$, and L_n are the standard Laguerre polynomials.

Problem 12.10. *If $B = \frac{1}{2}$, then the two terms of Eq. (12.14) become linearly dependent; i.e., they are proportional to each other. Since a second order differential equation must have two linearly independent solutions, show in this case a correct wave function is given by $\psi^{(-)}(r, \alpha) \approx \sqrt{r} (c_1 + c_2 \log r)$ both of whose terms are normalizable.*

Chapter 13

WKB and Supersymmetric WKB

In this chapter, we will study the WKB (Wentzel-Kramers-Brillouin)[1] method, which is one way to obtain approximate solutions for which exact solutions do not exist. We will then study the supersymmetric variant of the method, SWKB. We will discover the peculiar fact that the SWKB method yields exact solutions for all of the known translational shape invariant potentials.

13.1 WKB

This is a *semi-classical* approximation method; that is, it combines principles from both classical and quantum mechanics. In this case, approximate solutions to the Schrödinger equation are linked with the classical turning points of a particle; viz., the region in which a classical particle is constrained, where a quantum mechanical wave function is not. An example is the harmonic oscillator, for which a classical particle trajectory cannot exceed the amplitude of the oscillator, but the quantum mechanical wave function can and does.

We begin with the behavior of a particle in regions where the potential V is constant, such as a bound state of a finite square well. The time independent wave function is $\psi(x) = Ae^{\pm ikx}, E > V$; viz., inside the well, or $\psi(x) = Ae^{\pm \kappa x}, E < V$; viz., in the walls, where $k = \sqrt{E - V}$ and $\kappa = \sqrt{V - E}$. Let us confine ourselves to the former. This means that we are in the so-called classical region, whose limits on x are given by the solutions of $E = V(x)$. We now modify V to allow it to vary slowly in comparison to the variation in the wave function, characterized by k.

---

[1] Harold Jeffreys actually discovered this method in 1923, and is often credited by including his initial: JWKB.

(To put it another way, the "wavelength" representing the change in the magnitude of $V(x)$ is much longer than the wavelength $\lambda = 2\pi/k$ of the wave function.) In so doing, we expect A and k to be slowly varying functions of x.

From the definition of the classical momentum, $p(x) \equiv \sqrt{E - V(x)}$ and the quantum mechanical relationship $p \equiv -i d/dx$, the Schrödinger equation becomes

$$\frac{d^2\psi}{dx^2} = -p(x)^2\psi.$$

The wave function may then be written in terms of amplitude and phase:

$$\psi(x) = A(x)e^{i\phi(x)}.$$

Our goal is to find $A(x)$ and $\phi(x)$.

The Schrödinger equation now becomes:

$$\frac{d^2\psi}{dx^2} = [A'' + 2iA'\phi' + iA\phi'' - A(\phi')^2]e^{i\phi} = -p^2 A(x)e^{i\phi(x)} .$$

We have suppressed the x-dependence of p and, as usual, set $\hbar$ and $2m$ equal to 1.

Canceling $e^{i\phi}$,

$$A'' + 2iA'\phi' + iA\phi'' - A(\phi')^2 = -p^2 A. \qquad (13.1)$$

Separating real and imaginary parts ,

$$A'' - A(\phi')^2 = -p^2 A. \qquad (13.2)$$

$$2A'\phi' + A\phi'' = 0. \qquad (13.3)$$

The first of these can be written as

$$A'' = A[(\phi')^2 - p^2]. \qquad (13.4)$$

The second can be written as

$$(A^2\phi')' = 0. \qquad (13.5)$$

Integrating Eq. (13.5) yields the exact solution for $A(\phi)$:

$$A = \frac{C}{\sqrt{\phi'}}.$$

Equation (13.4) has no exact solution, so it is here that the approximation must be made. Since we have set the condition that $A(x)$ must vary slowly compared to the oscillations of the wave function, its second derivative is small: $A''/A \simeq 0$. Then $(\phi')^2 \simeq p^2$. Treating this as an equality, $\phi' = \pm p$. Let us integrate this, taking as our limits x_0 and x and restoring the notation with the x-dependence of p.

$$\phi(x) - \phi(x_0) = \pm \int_{x_0}^{x} p(x')dx'.$$

We can set $\phi(x_0) = 0$ and choose $x_0 = 0$, whence $\phi(x) = \pm \int_0^x p(x')dx'$. So the approximate wave function is

$$\psi(x) \simeq Ae^{i\phi} \simeq \frac{C}{\sqrt{p(x)}} e^{\pm i \int_0^x p(x')dx'}. \tag{13.6}$$

This is the general WKB result. Note that we have used the classical momentum $p = \sqrt{E - V}$, but the wave function is non-zero beyond the classical turning points, where $E < V$ and p is thus imaginary. In those regions, $p(x) \to | p(x) |$ and $e^{\pm i \int_0^x p(x')dx'} \to e^{\pm \int_0^x |p(x')|dx'}$.

Let us apply the WKB method to an arbitrarily shaped potential $V(x)$ added to the bottom of an infinite square well whose potential is taken as 0, in the range $0 \le x \le a$ as illustrated in Figure 13.1.

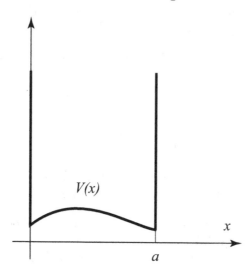

Fig. 13.1   Arbitrarily shaped potential $V(x)$ added to the bottom of an infinite square well.

For the time being, let us consider those regions where $E > V(x)$. Then ψ is in general a linear combination of terms involving $\pm i \int_0^x p(x')dx'$. We can rewrite it as

$$\psi(x) \simeq \frac{1}{\sqrt{p(x)}} \left[C_1 \sin\left(\int_0^x p(x')dx' \right) + C_2 \cos\left(\int_0^x p(x')dx' \right) \right].$$

Using the boundary conditions for infinite vertical walls, $\psi(0) = 0 \to C_2 = 0$ and $\psi(a) = 0 \to$

$$\int_0^a p(x)dx = n\pi \tag{13.7}$$

In the regions where $E < V(x)$, $\psi(x) \sim \pm \int_0^x |p(x')| \, dx'$ or

$$\psi(x) \simeq \frac{1}{\sqrt{|p(x)|}} \left[C_1 \sinh\left(\int_0^x |p(x')|dx' \right) + C_2 \cosh\left(\int_0^x |p(x')|dx' \right) \right].$$

Let us note some interesting and important features of this result. First, the form of Eq.(13.7) is reminiscent of the Bohr-Sommerfeld condition from pre-Schrödinger quantum theory: this is what makes the WKB method semi-classical. Second, knowing the form of $V(x)$ immediately gives us the approximate energy levels E_n^{WKB}: $\int_0^a p(x)dx = \int_0^a \sqrt{E - V(x)}dx = n\pi$. Third, if we remove the potential $V(x)$, we should obtain the exact solution for the infinite well: $V(x) = 0 \to p(x) = \sqrt{E}$, a constant. Then $n\pi = \int_0^a p(x)dx = pa = \sqrt{E}a$ gives $E = n^2\pi^2/a^2$.

Problem 13.1. *An infinite well has a square barrier of height V_0 and width $a/3$ in its middle:*

$$V(x) = \begin{cases} 0, & 0 < x < a/3 \\ V_0, & a/3 < x < 2a/3 \\ 0, & 2a/3 < x < a \\ \infty & \text{elsewhere} \end{cases}.$$

For $E > V_0$, use the WKB method to find the approximate value of energy level E_n in terms of V_0 and the infinite well energy $E_n^0 = (n\pi/a)^2$.

For $E < V_0$,; viz., within the barrier, the WKB method can obtain the tunneling probability. We shall not do this here[2].

The condition $\int_0^a p(x)dx = n\pi$ is specific to cases with two infinite vertical walls. The factor on the right hand side will change for cases with one vertical wall, such as a half harmonic oscillator with bumps, or no vertical walls, such as a full harmonic oscillator with bumps. For these cases, the analysis gets more complicated. The reason is that for any finite

---

[2]See A. Messiah, *Quantum Mechanics*, v.1, Paris: Dunod Publ. (1964).

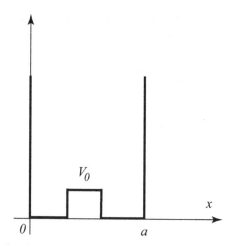

Fig. 13.2 The potential for problem (13.1) .

potential, $E \simeq V$ at the classical turning points. At these points, p goes to zero, so the WKB wave function $\psi \sim \frac{1}{\sqrt{p}}$ blows up.

To resolve this dilemma, we need to make another approximation. Since the WKB wave function is already approximate, we can modify it in some small way to avoid the turning-point catastrophe. This is usually done by a "patching" technique. Instead of the WKB wave functions to the right and left of the well boundary matching exactly (the source of the problem) they are spliced to a simpler patching function ψ_p which traverses the boundary. To be certain that this is not severely changing the WKB results, we insist that the patching function only be valid in a region very close to the turning points. The usual way to do this is to linearize the potential in those regions. Choosing for the moment $x = 0$ at the right hand turning point, $V(x) \simeq E + V'(0)x$. The Schrödinger equation is then

$$-\frac{d^2\psi_p}{dx^2} + [E + V'(0)x]\psi_p = E\psi_p.$$

Two substitutions simplify this. Defining $\alpha \equiv \sqrt[3]{V'(0)}$ and $z \equiv \alpha x$ yields

$$\frac{d^2\psi_p}{dz^2} = z\psi_p. \tag{13.8}$$

This is the Airy Equation, whose solutions are the Airy functions. They are quite complicated (which is why the linear potential problem is rarely seen in introductory quantum mechanics). We shall not go into any derivation of the connection formulas which splice these functions onto the WKB wave functions [3]. We simply state the results. For x_1 the left hand turning point and x_2 the right hand turning point of a potential well,

$$
\psi(x) \simeq
\begin{cases}
\dfrac{C_3}{\sqrt{|p(x)|}}\, e^{-\int_x^{x_1} |p(x')|dx'}, & x < x_1 \\[3mm]
\dfrac{2C_3}{\sqrt{p(x)}}\, \sin\left[\int_{x_1}^x p(x')dx' + \pi/4\right], & x > x_1
\end{cases}
\tag{13.9}
$$

$$
\psi(x) \simeq
\begin{cases}
\dfrac{C_4}{\sqrt{|p(x)|}}\, e^{-\int_{x_2}^x |p(x')|dx'}, & x > x_2 \\[3mm]
\dfrac{2C_4}{\sqrt{p(x)}}\, \sin\left[\int_x^{x_2} p(x')dx' + \pi/4\right], & x < x_2
\end{cases}
\tag{13.10}
$$

(We could find C_3 and C_4 by normalizing the wave functions, but we don't need them.)

We are now ready to find the the quantization condition for $\int p(x)dx$. For the case of one vertical wall at $x_1 \equiv 0$, $\psi(0) = 0$. Since we are in the region within the turning points $x < x_2$, the second line of Eq. (13.10) with $x = 0$ makes the argument of the sine a multiple of π: $\int_0^{x_2} p(x)dx + \pi/4 = n\pi$. Then the quantization condition for one vertical wall is

$$
\int_0^{x_2} p(x)dx = (n - 1/4)\pi, \, n = 1, 2, 3, \dots .
\tag{13.11}
$$

For no vertical walls, we compare the solutions of Eq. (13.9) and Eq. (13.10) for $x_1 < x < x_2$. They must agree because they cover the same region. I.e.;

$$
C_3 \sin\left[\int_{x_1}^x p(x')dx' + \pi/4\right] = C_4 \sin\left[\int_x^{x_2} p(x')dx' + \pi/4\right] .
$$

This is only possible if the constants C_3 and C_4 are the same magnitude and the sines are the same. The latter are non-trivially the same if their arguments differ by some multiple of π.

Since any minus sign in the argument of the sine can be absorbed into the constant, that multiple is $n\pi$, not, as we might have thought, $2n\pi$. So in fact the constants are related by $C_4 = \pm C_3$. Thus

$$
\int_x^{x_2} p(x')dx' + \pi/4 = \pm\left[\int_{x_1}^x p(x')dx' + \pi/4\right] + n\pi.
$$

---

[3] For these details, see e.g. D. J. Griffiths, *Quantum Mechanics*, 2nd ed., CA: Benjamin-Cummings (2004).

For WKB quantization we need $\int_{x_1}^{x_2} p(x')dx'$. For the plus sign, we obtain

$$\int_x^{x_2} p(x')dx' + \pi/4 = + \left[\int_{x_1}^x p(x')dx' + \pi/4 \right] + n\pi.$$

Then

$$\int_x^{x_2} p(x')dx' - \int_{x_1}^x p(x')dx' = n\pi.$$

But this does not give the right integral for quantization. On the other hand, for the minus sign, we obtain

$$\int_x^{x_2} p(x')dx' + \pi/4 = - \left[\int_{x_1}^x p(x')dx' + \pi/4 \right] + n\pi.$$

This gives us the appropriate integration:

$$\int_{x_1}^x p(x')dx' + \int_x^{x_2} p(x')dx' = \int_{x_1}^{x_2} p(x')dx' = (n - 1/2)\pi, \quad n = 1, 2, \ldots .$$

$$(13.12)$$

We have now obtained the WKB conditions for potentials with two, one, and zero vertical walls. For the first, we have done the example of the infinite square well and obtained the exact energies, despite the fact that the WKB method is an approximate one. There are several other potentials for which the WKB method give exact results. Let us now solve two of them: the half-harmonic oscillator (one vertical wall) and the full harmonic oscillator (no vertical walls).

Problem 13.2. *The half-harmonic oscillator potential is*

$$V(x) = \begin{cases} \infty, & x = 0 \\ \frac{1}{4}\omega^2 x^2, & x > 0 \end{cases}.$$

Show that the WKB approximation gives the exact result for the energies [4].

Problem 13.3. *The harmonic oscillator potential is* $V(x) = \frac{\omega^2 x^2}{4}, -\infty < x < \infty$. *Show that the WKB approximation gives the exact result for the energies.*

The infinite well results are not hard to explain, once you know the answer. The wave functions are $\psi_n(x) = A\sin(n\pi x/a)$, where A is constant. Then $A'' = 0$ in Eq. (13.4), rendering the WKB "approximation" exact. The exactness of WKB for the half-and full harmonic oscillators has not yet been explained.

---

[4]Keep in mind that the usual notation starts with $n = 0$, while the WKB notation starts with $n = 1$.

13.2   SWKB

In Chapter 6 we introduced shape invariance and described the method by which we could obtain the eigenvalues and eigenfunctions of translational shape invariant potentials. Surprisingly, the exact energies for shape invariant potentials can be obtained from the SWKB method. Since any supersymmetric potential is given in terms of the superpotential W by $V = W^2 - W'$, the ordinary WKB approximation is

$$\int_{x_1}^{x_2} p(x)dx = \int_{x_1}^{x_2} \sqrt{E - V(x)}dx' = \int_{x_1}^{x_2} \sqrt{E - (W^2 - W')}dx = \text{N}\pi$$

$$(13.13)$$

where $\text{N} = n, n - 1/4, n - 1/2$ for two, one and no vertical walls respectively. This formulation cannot in general give exact results, since it is the WKB approximation itself. But if W' is removed, then $\int_{x_1}^{x_2} \sqrt{E - W^2}dx = n\pi$ for integer n gives the exact energies for all of the known translational shape invariant potentials. Let us emphasize that

(1) no one has yet explained why this is so, although much work has been done and progress has been made;
(2) we are not sure that we have found all such potentials;
(3) the method does not work for potentials other than those which have translational shape invariance. To put it another way, *if* a potential belongs to the family of known translational shape invariant potentials, *then* we can find its exact energies from the SWKB formulation, but as yet we do not know if the converse is true.

13.2.1   *Infinite Square Well: SWKB*

As an example, let us solve the infinite square well problem. As we saw in Chapter 5 we began with the superpotential $W(x) = -b\cot x$ with the parameter $b > 0$, where we chose the domain $0 < x < \pi$. We found the supersymmetric partner potentials to be:

$$V_-(x, b) = W^2(x) - \frac{dW}{dx} = b(b - 1)\,\text{cosec}^2 x - b^2$$

$$\text{and} \qquad\qquad (13.14)$$

$$V_+(x, b) = W^2(x) + \frac{dW}{dx} = b(b + 1)\,\text{cosec}^2 x - b^2 \ .$$

For $b = 1$, the potential $V_-(x, 1)$ was an infinite one-dimensional square well potential whose bottom was set to -1. Then the SWKB equation is

$$J(E) = \int_{-\cot^{-1}\sqrt{E}}^{\cot^{-1}\sqrt{E}} \sqrt{E - \cot^2 x}\, dx = n\pi. \qquad (13.15)$$

As we integrate this expression, we will demonstrate techniques that are often used in physics to solve complicated integrals. We first make a substitution $x = \frac{\pi}{2} + y$. This leads to

$$J(E) = \int_{-\tan^{-1}\sqrt{E}}^{\tan^{-1}\sqrt{E}} \sqrt{E - \tan^2 y}\, dy. \qquad (13.16)$$

To solve it further, we introduce a parameter α which we will later set equal to one. We define

$$I(\alpha, E) = \int_{-\tan^{-1}\sqrt{E}}^{\tan^{-1}\sqrt{E}} \sqrt{E - \alpha \cdot \tan^2 y}\, dy. \qquad (13.17)$$

Thus, $J(E) = I(1, E)$. Now, differentiating $I(\alpha, E)$ with respect to α, we obtain

$$\frac{\partial I(\alpha)}{\partial \alpha} = -\frac{1}{2} \int_{-\tan^{-1}\sqrt{E}}^{\tan^{-1}\sqrt{E}} \left(\frac{\tan^2 y}{\sqrt{E - \alpha \cdot \tan^2 y}} \right) dy$$

$$= -\frac{1}{2\sqrt{\alpha}} \sin^{-1}\left(\sqrt{\frac{\alpha}{E}}\, \tan y \right)$$

$$+ \frac{1}{2\sqrt{E + \alpha}} \sin^{-1}\left(\sqrt{\frac{E + \alpha}{E}}\, \tan y \right). \qquad (13.18)$$

Problem 13.4. *Derive the above expression for* $\frac{\partial I(\alpha)}{\partial \alpha}$.

This implies that

$$I(\alpha, E) = \int d\alpha \left[-\frac{1}{2\sqrt{\alpha}} \sin^{-1}\left(\sqrt{\frac{\alpha}{E}}\, \tan y \right) \right. \qquad (13.19)$$

$$\left. + \frac{1}{2\sqrt{E + \alpha}} \sin^{-1}\left(\sqrt{\frac{E + \alpha}{E}}\, \sin y \right) \right].$$

In the first integral, setting $\sqrt{\alpha} = \beta$, we obtain

$$\int d\alpha \left[\frac{1}{\sqrt{\alpha}} \sin^{-1}\left(\sqrt{\frac{\alpha}{E}}\, \tan y \right) \right] = \int 2\, d\beta\, \sin^{-1}\left(\frac{\beta}{\sqrt{E}}\, \tan y \right) \equiv T(\beta). \qquad (13.20)$$

Now using $\int dz \sin^{-1} z = z \sin^{-1} z + \sqrt{1 - z^2}$, we obtain

$$T(\beta) = \int 2\frac{\sqrt{E}}{\beta} \, d\left(\frac{\beta}{\sqrt{E}} \tan y\right) \, \sin^{-1}\left(\frac{\beta}{\sqrt{E}} \tan y\right)$$

$$= 2\beta \sin^{-1}\left(\frac{\beta}{\sqrt{E}} \tan y\right) + 2\sqrt{1 - \beta^2 \frac{\tan^2 y}{E}} \frac{\sqrt{E}}{\tan y} .$$

Similarly integrating $K(\alpha) = \int d\alpha \, \dfrac{\sin^{-1}\left(\sqrt{\frac{E+\alpha}{E}} \sin y\right)}{\sqrt{E+\alpha}}$, we obtain

$$K(\alpha) = 2\sqrt{E+\alpha} \sin^{-1}\left(\frac{\sqrt{E+\alpha}}{\sqrt{E}} \sin y\right) + 2\frac{\sqrt{E}}{\sin y}\sqrt{1 - (E+\alpha)\frac{\sin^2 y}{E}},$$

where $\gamma = \sqrt{E+\alpha}$. Now, collecting terms, we obtain

$$I(\alpha, E) = \sqrt{E+\alpha} \sin^{-1}\left(\frac{\sqrt{E+\alpha}}{\sqrt{E}} \sin y\right) - \sqrt{\alpha} \sin^{-1}\left(\sqrt{\frac{\alpha}{E}} \tan y\right) .$$

Thus,

$$J(E) = I(1, E) = 2 \int_0^{\tan^{-1}\sqrt{E}} \sqrt{E - \tan^2 y} \, dy$$

$$= 2 \left[\sqrt{E+1} \, \frac{\pi}{2} - \frac{\pi}{2}\right] = \left[\sqrt{E+1} - 1\right]\pi .$$

$$(13.21)$$

Now, setting $\left[\sqrt{E+1} - 1\right]\pi = n\pi$, we obtain $E_n = (n+1)^2 - 1$. For $n = 0$, we obtain $E = 0$ as expected for unbroken susy.

Problem 13.5. *Obtain the energies for the harmonic oscillator from the SWKB method.*

13.2.2 Morse Potential: WKB and SWKB

We will solve for the energies using the WKB method, for which it is exact, then leave the SWKB method for a problem. The Morse superpotential is

$$W(x, A, B) = A\left(1 - e^{-z}\right) . \tag{13.22}$$

The WKB condition for Morse is given by

$$\int \left[E - A^2\left(1 - e^{-z}\right)^2 + Ae^{-z}\right]^{\frac{1}{2}} dz = n\pi . \tag{13.23}$$

Let us do a change of variable $e^{-z} = y$, the left hand side of Eq. (13.23) becomes

$$-A \int \frac{dy}{y} \left[\underbrace{\left(\frac{E + A + \frac{1}{4}}{A^2} \right)}_{\beta^2} - \left\{ (1 - y) + \frac{1}{2A} \right\}^2 \right]^{\frac{1}{2}} .$$

With two more change of variables: $(1 - y) + \frac{1}{2A} = u$ and $u = \beta \sin \theta$, the above integral becomes

$$A \beta^2 \int_{-\pi/2}^{\pi/2} \left(\frac{\cos^2 \theta}{1 - \beta \sin \theta + \frac{1}{2A}} \right) d\theta .$$

An integration then leads to

$$\frac{1}{2A\beta^2} \left[2\sqrt{-4A^2 (\beta^2 - 1) + 4A + 1} \tan^{-1} \left(\frac{2A\beta - (2A + 1) \tan \left(\frac{\theta}{2} \right)}{\sqrt{-4A^2 (\beta^2 - 1) + 4A + 1}} \right) \right]$$
$$+ \left[\frac{-2A\beta \cos(\theta) + 2A\theta + \theta}{2A\beta^2} \right] .$$

(13.24)

Setting $-4A^2 (\beta^2 - 1) + 4A + 1 = 4A^2(1 - \alpha^2)$, where $\alpha = \frac{\sqrt{E}}{A}$, we obtain

$$\frac{4A\sqrt{1 - \alpha^2} \tan^{-1} \left(\frac{2A\beta - (2A+1) \tan \left(\frac{\theta}{2} \right)}{2A\sqrt{1 - \alpha^2}} \right) - 2A\beta \cos(\theta) + 2A\theta + \theta}{2A\beta^2} . \quad (13.25)$$

Substituting the limits for θ; i.e., $\pm \pi/2$,

$$\frac{1}{2} \left(4A\sqrt{1 - \alpha^2}(-\pi/2) + (2A + 1)\pi \right) = \left(n + \frac{1}{2} \right) \pi .$$

(13.26)

Solving the above equation for energy, we obtain

$$E_n = A^2 - (A - n)^2 , \quad (13.27)$$

the exact eigenvalues for Morse potential.

Problem 13.6. *Fill in the steps from Eq. (13.23) to Eq. (13.27).*

Problem 13.7. *Obtain the energy spectrum for the Morse potential using the SWKB method.*

Let us make some concluding remarks. First, there is no evidence that SUSYQM is exact for solvable potentials other than those which are translationally shape invariant.

Second, if we put $\hbar$ back in, it multiplies W' but not W: $V_- = W^2 - \hbar W'$. Then the square root in the WKB integral may be expanded in powers of $\hbar$. That is why some papers and books refer to the SWKB integral as the lowest order approximation. To order $\hbar^6$ it is known that the terms are each 0 [5]. No general proof to all orders yet exists.

Third, we have seen several cases for which the WKB method gives exact results: the infinite well and Morse potentials, the reasons for which are understood, and the half- and full-harmonic oscillator potentials which are not understood. In the case of the Coulomb potential, it is also possible for the WKB method to give exact results, but only if the centrifugal term $\frac{\ell(\ell+1)}{r^2}$ is replaced by $\frac{(\ell+1/2)^2}{r^2}$ This is called the Langer correction [6], and is also not yet understood.

There is much work to be done to account for the exactness of the SWKB method for all known shape invariant potentials. If other such potentials should be found, they will be subjected to this test.

---

[5] R. Adhikari, R. Dutt, A. Khare, and U. Sukhatme, "Higher order WKB approximations in supersymmetric quantum mechanics", *Phys. Rev.* **A38**, 1679 (1986).

[6] R. E. Langer, "On the connection formulas and solutions of the wave equation", *Phys. Rev.* **51**, 669 (1937).

Chapter 14

The Quantum Hamilton-Jacobi Formalism and SUSYQM

In this chapter we introduce another formulation of quantum mechanics, the Quantum Hamilton-Jacobi (QHJ) formalism [1] and its connection to SUSYQM. In this formalism, we work with the quantum momentum function $p(x)$, which is the quantum analog of the classical momentum function $p_c(x) = \sqrt{E - V}$.

In QHJ formalism the spectrum of a quantum mechanical system is determined by the solution of the following non-linear differential equation:

$$p^2 - i p' = E - V(x, \alpha) \equiv p_c^2 \ . \tag{14.1}$$

This equation is related to the Schrödinger equation

$$-\psi'' + (V(x, \alpha) - E)\psi = 0 \tag{14.2}$$

via the correspondence

$$p(x) = -i\psi'(x)/\psi(x) \quad \text{whence} \quad \psi(x) \sim e^{-\int p(x)\,dx} \ . \tag{14.3}$$

The quantum momentum function (QMF) is clearly a function of x, E, and parameter α; i.e., $p \equiv p(x, E, \alpha)$, however we will suppress this dependence unless necessary. From Eq. (14.2), we see that function p for the zero-energy ground state equals $iW(x, \alpha)$, the superpotential of the system. For non-zero energies, the QMF has singularities at the zeroes of the wave function $\psi(x)$. The structure of these singularities plays an important role in determining the spectrum of the hamiltonian. In Eq. (14.1), note that if the potential is a non-singular function, the singularities of the p^2, if any, would have to be canceled by those of the term p'. We will see that Eq. (14.1) helps us determine the singularities of the p-function.

---

[1] Even though this procedure has similarities with the Hamilton-Jacobi formalism of classical mechanics, the material covered in this chapter is self contained.

The Quantum Hamilton-Jacobi method is centered around analyzing the pole structure of $p = -i\left(\frac{\psi'}{\psi}\right)$. Since the n-th eigenfunction of the hamiltonian has n nodes within the turning points determined by the energy E_n and the potential, the corresponding function p will have n singular points. We will now show that these singularities are simple poles; i.e., near a singular point x_0, the function has the form $p \approx \frac{\delta}{x-x_0}$, where the constant δ is known as the residue.

Near a point x_0 where the wave function $\psi(x)$ is zero, we can write the wave function as $\psi(x) = h(x)(x - x_0)$. Its derivative is then given by $\psi'(x) = h'(x)(x - x_0) + h(x)$. The singular term of the QMF is then given by

$$p(x) \equiv -i\,\psi'(x)/\psi(x) = \frac{-i}{x - x_0} \ . \tag{14.4}$$

I.e., the p has a pole at x_0 with a residue of $-i$. Since these poles are due to the zeroes of the wave function and their location depends on energy, they are defined as the "moving poles." In addition to these moving poles, QMF may also have a poles at the singular points of the *potential*. Since these are fixed for a given system described by the potential V, they are known as the "fixed" poles. Now let us visualize the x-axis to be the real axis of a complex plane. These moving poles would then all fall on the real axis. If we consider a closed path that encircles these moving poles, and there are n of them, the Cauchy theorem would imply that the line integral of p carried out on this closed path traveling in a counter clockwise direction would yield:

$$\frac{1}{2\pi} \oint p(x)\, dx = n \tag{14.5}$$

where contour involved includes all n moving poles of the system and none of the fixed poles. In complex analysis, we call it a contour integration.

The strength of QHJ derives from the fact that the contour of the above integration, which was put in to encircle moving poles only, can be deformed to enclose the fixed poles instead (albeit traveling in the opposite direction.) To understand this, visualize the complex plane folded into a sphere where the boundary at infinity is pulled together to a point. Now, exactly as a rubber-band around Australia on a globe can be stretched to encircle all nations other than Australia, similarly a contour on a complex plane around moving poles can be deformed to enclose all fixed poles instead. Figures 14.1(a) through 14.1(f) show such a deformation. Thus, $\oint p(x)\, dx\big|_{\text{fixed poles}} = -\oint p(x)\, dx\big|_{\text{moving poles}}$. The difference in sign is due

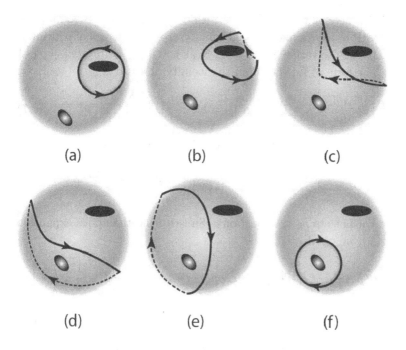

Fig. 14.1   Continuously moving a contour on complex plane

to the change in direction in which we move along the contour as it is deformed. The counter-clockwise becomes clockwise as shown in Figure 14.1. Thus, the condition of Eq. (14.5) becomes

$$\frac{1}{2\pi} \oint p\, dx \bigg|_{\text{fixed poles}} = -n \ . \tag{14.6}$$

In QHJ, we first determine the singularity structure of p from Eq. (14.1) by explicitly inserting the potential V. Then Eq. (14.6) determines n as a function of the energy E_n and parameter α. Inverting this equation, we find the energy $E_n(\alpha)$.

Since Eq. (14.5) determines exact eigenvalues of the system, it is known as the quantization condition. It is important to note that unlike semi-classical quantization discussed in Chapter 13, the condition Eq. (14.5) is exact.

As an example, let us consider the problem of a harmonic oscillator described by the potential $V(x, \omega) = \frac{1}{4} \omega x^2$. The corresponding super-potential is given by $W(x, \omega) = \frac{1}{2} \omega x$. At first sight, this seems to be a potential with no singularity. However, it does diverge as $x \to \infty$ and will force the QMF to have a fixed pole at infinity. To see the form of p at infinity let us change the variable to $u = \frac{1}{x}$. Near $u = 0$, we write $p(x) = p\left(\frac{1}{u}\right)$ as $\widetilde{p}(u,)$. The form of Eq. (14.1) in the vicinity of $u = 0$ is

$$\widetilde{p}^2 + iu^2 \frac{d\widetilde{p}}{du} = E - V\left(\frac{1}{u}, \alpha\right) = E - \frac{1}{4}\,\omega\left(\frac{1}{u}\right)^2 = p_c^2 . \quad (14.7)$$

We expand the quantum momentum function $\widetilde{p}(u)$ in powers of u:

$$\widetilde{p}(u, \alpha_0) = \sum_{k=-1}^{\infty} \widetilde{p}_k\, u^k = \frac{\widetilde{p}_{-1}}{u} + \widetilde{p}_0\, u^0 + \widetilde{p}_1\, u^1 + \widetilde{p}_2\, u^2 + \cdots$$

where the coefficients $\widetilde{p}_k$ are constants. Substituting the above expansion in Eq. (14.7), we obtain

$$\widetilde{p}_{-1}^2 u^{-2} + \left(2\,\widetilde{p}_{-1}\widetilde{p}_0\right) u^{-1} + \left(\widetilde{p}_0^2 + 2\widetilde{p}_{-1}\widetilde{p}_1 - i\widetilde{p}_{-1}\right) + \cdots = E - \frac{1}{4}\,\omega\left(\frac{1}{u}\right)^2 .$$

$$(14.8)$$

By identifying the various powers of the variable u, we obtain

$$(\widetilde{p}_{-1})^2 = -\frac{1}{4}\,\omega^2 ; \quad \widetilde{p}_0 = 0; \quad \text{and} \quad 2\widetilde{p}_{-1}\widetilde{p}_1 - i\widetilde{p}_{-1} = E . \quad (14.9)$$

The first equality leads to $\widetilde{p}_{-1} = \pm\frac{1}{2}i\omega$. At this point, we choose $\widetilde{p}_{-1} = \frac{1}{2}i\omega$ as the correct root. This can be justified from the requirement that for $E \to 0$, the QMF should become iW. Thus, in this QHJ formulation we make our first connection to supersymmetric quantum mechanics[2]. This choice then leads to $i\omega\widetilde{p}_1 + \frac{1}{2}\omega = E$. From the last equality, we obtain $\widetilde{p}_1 = -i\left(\frac{E}{\omega} - \frac{1}{2}\right)$. As we will soon see, this last coefficient is the only one that contributes to the contour integration around the fixed pole for this system.

The quantization condition from Eq. (14.6) yields

$$-n = \frac{1}{2\pi} \oint p(x, E_n, \alpha)\, dx\Big|_{\text{fixed poles}} .$$

---

[2] For an alternate explanation of this choice without resorting to SUSYQM, see R. A. Leacock and M. J. Padgett, "HamiltonJacobi/action-angle quantum mechanics", *Phys. Rev.* **D28**, 2491-2502, (1983.) However, we find the stipulation $\lim_{E \to 0} p \to iW$ to be a much easier method to unambiguously select the correct root. In our case that implies $\widetilde{p}_{-1}/u = \frac{1}{2}i\omega/u$. Hence, $\widetilde{p}_{-1} = \frac{1}{2}i\omega$.

Since the fixed pole is at infinity, we change variable to $u = 1/x$ and replace dx by $-du/u^2$. This then gives

$$n = \frac{1}{2\pi} \oint_{u=0} \frac{1}{u^2} \, \widetilde{p} \, du$$

$$= \frac{1}{2\pi} \oint_{u=0} \frac{1}{u^2} \left(\frac{\widetilde{p}_{-1}}{u} + \widetilde{p}_0 \, u^0 + \widetilde{p}_1 \, u^1 + \widetilde{p}_2 \, u^2 + \cdots \right) du$$

$$= \frac{1}{2\pi} \oint_{u=0} \left(\frac{\widetilde{p}_{-1}}{u^3} + \frac{\widetilde{p}_0}{u^2} + \frac{\widetilde{p}_1}{u} + \widetilde{p}_2 \, u^0 + \cdots \right) du$$

$$= i\widetilde{p}_1 \, .$$

This then yields $n = i\widetilde{p}_1 = \left(\frac{E_n}{\omega} - \frac{1}{2} \right)$. I.e., $E_n = \left(n + \frac{1}{2} \right) \omega$, the familiar result for the harmonic oscillator.

14.1 Connection to SUSYQM and Shape Invariance

We have already seen how SUSYQM helped us choose the correct root in QHJ formalism with relative ease. In this section we will work out the general connection between SUSYQM and shape invariance and QHJ theory. Since parameters play a very important role in shape invariance, we will now explicitly display all variables that the QMF p and potential V depend upon. Since the partner potentials $V_\pm(x, \alpha)$ are given by $W^2(x, \alpha) \pm W'(x, \alpha)$, the QHJ equation (14.1) for the hamiltonian $H_-(x, \alpha)$ becomes

$$p^{(-)\,2}(x, E^{(-)}, \alpha) - i\,p^{(-)\,\prime}(x, E^{(-)}, \alpha) = E^{(-)} - \left[W^2(x, \alpha) - W'(x, \alpha) \right] \, . \tag{14.10}$$

Thus, in the limit[3] $E^{(-)} \to 0$, we have

$$\lim_{E \to 0} p^{(-)}(x, E^{(-)}, \alpha) \to i\,W(x, \alpha) \, . \tag{14.11}$$

Considering now the supersymmetric partner hamiltonian $H_+(x, \alpha)$, there exists another analogous equation for a partner QMF, $p^{(+)}(x, \alpha)$ given by

$$p^{(+)\,2}(x, E^{(+)}, \alpha) - i\,p^{(+)\,\prime}(x, E^{(+)}, \alpha) = E^{(+)} - \left[W^2(x, \alpha) + W'(x, \alpha) \right] \, . \tag{14.12}$$

---

[3]We will assume that our superpotential is such that supersymmetry remains unbroken for a range of values of the parameters α.

Since supersymmetry ensures that these equations lead to the same set of eigenvalues, except for the ground state, let us write $E \equiv E^{(-)} = E^{(+)}$. The corresponding Schrödinger equations are

$$H_{\pm}(x,\alpha)\,\psi^{(\pm)} = -\psi^{(\pm)\,\prime\prime} + \left[W^2(x,\alpha) \pm W'(x,\alpha)\right]\psi^{(\pm)} = E\,\psi^{(\pm)}\,,$$
$$(14.13)$$

and the solutions are connected to the QHJ solutions by

$$p^{(\pm)} = -i\left(\frac{\psi^{(\pm)\,\prime}}{\psi^{(\pm)}}\right)\,. \qquad (14.14)$$

Defining operators $A^+ = -\frac{d}{dx} + W$ and $A^- = \frac{d}{dx} + W$, we can rewrite the partner hamiltonians as $H_+ = A^- A^+$ and respectively $H_- = A^+ A^-$. Furthermore, we find that A^+ and A^- behave as raising and lowering operators between the eigenstates of the partner hamiltonians. In particular,

$$\psi^{(-)} = C^{(-)}A^+\psi^{(+)} = C^{(-)}\left(-\psi^{(+)\,\prime} + W\,\psi^{(+)}\right)$$
$$\psi^{(+)} = C^{(+)}A^-\,\psi^{(-)} = C^{(+)}\left(\psi^{(-)\,\prime} + W\,\psi^{(-)}\right)\,, \qquad (14.15)$$

where $C^{(-)}$ and $C^{(+)}$ are normalization constants. To find a relationship between $p^{(-)}$ and $p^{(+)}$ we exploit the connection between $\psi^{(-)}$ and $\psi^{(+)}$ respectively. Using Eqs. (14.13), (14.14), (14.15), we obtain

$$\frac{\psi^{(+)\,\prime}}{\psi^{(-)}} = C^{(+)}\left(W^2 - E + i\,Wp^{(-)}\right) \qquad (14.16)$$

$$\frac{\psi^{(-)\,\prime}}{\psi^{(+)}} = C^{(-)}\left(-W^2 + E + i\,Wp^{(+)}\right)\,. \qquad (14.17)$$

Multiplying Eqs. (14.16) and (14.17) we obtain

$$-p^{(+)}p^{(-)} = C^{(-)}C^{(+)}\left(W^2 - E + i\,W\,p^{(-)}\right)\left(-W^2 + E + i\,Wp^{(+)}\right)\,.$$

To solve the above equation for $p^{(+)}$ or $p^{(-)}$, we first evaluate the product of the normalization constants $C^{(-)}C^{(+)}$. Since $\psi^{(+)} = C^{(+)}A^-\psi^{(-)}$, and $\psi^{(-)} = C^{(-)}A^+\psi^{(+)}$, we obtain successively

$$\langle\psi^{(+)}|\psi^{(+)}\rangle = C^{(+)}\langle\psi^{(+)}|A^-|\psi^{(-)}\rangle$$
$$= C^{(+)}\langle\psi^{(+)}|A^-C^{(-)}A^+|\psi^{(+)}\rangle$$
$$= C^{(+)}C^{(-)}E\langle\psi^{(+)}|\psi^{(+)}\rangle\,, \qquad (14.18)$$

where we have used $A^- A^+ = H_+$ as noted earlier. Thus, $C^{(+)}C^{(-)} = 1/E$. This leads to

$$p^{(+)} = \frac{i\,Wp^{(-)} + W^2 - E}{-p^{(-)} + i\,W}\,, \qquad (14.19)$$

or,

$$p^{(-)} = \frac{-i\,W p^{(+)} + W^2 - E}{-p^{(+)} - i\,W} \ . \tag{14.20}$$

It is important to note at this point that both sides of Eqs. (14.19) and (14.20) are related to the same superpotential $W(x, \alpha)$ and can be denoted by $p^{(\pm)}(x, \alpha)$. Also note that $p^{(+)}$ in Eq. (14.19) is not defined for the ground state for which $-p^{(-)} + i\,W = 0$ [4].

We have only applied the conditions of supersymmetry so far. To render a hamiltonian solvable, the superpotential $W(x, \alpha)$ needs to satisfy the shape invariance condition, thus enabling us to find the solution of H_+ or H_- by algebraic means.

As we have seen before in Chapter 6, for shape invariant potentials $V^{(\pm)}(x, \alpha_i)$ we have $V^{(+)}(x, \alpha_i) = V^{(-)}(x, \alpha_{i+1}) + R(\alpha_i)$, where $R(\alpha_i)$ is a constant. Consequently, the eigenstates $\psi_n^{(+)}(x, \alpha_i)$ are identical to $\psi_n^{(-)}(x, \alpha_{i+1})$. In addition, since SUSYQM relates $\psi_n^{(+)}(x, \alpha_i)$ to $\psi_{n+1}^{(-)}(x, \alpha_i)$ via operators A and A^+, the states $\psi_{n+1}^{(-)}(x, \alpha_i)$ can be derived from states $\psi_n^{(-)}(x, \alpha_{i+1})$, and so forth. That is, a simple parameter shift allows for construction of the entire ladder of eigenstates $\psi_n^{(-)}$ or $\psi_n^{(+)}$. Furthermore, $E_n = \sum_{i=0}^{n-1} R(\alpha_i)$. We shall employ a similar technique to construct the ladders of QMF's $p^{(-)}$. Let us now replace the subscript n, which was a label for energy E, with E itself. Subscript $n - 1$ is replaced by $E - R(\alpha_i)$. Thus, shape invariance identifies $\psi_E^{(+)}(x, \alpha_i)$ with $\psi_{E-R(\alpha_i)}^{(-)}(x, \alpha_{i+1})$, which in QHJ becomes a relationship between quantum momentum functions. From Eq. (14.14), we find $p_{E-R(\alpha_i)}^{(-)}(x, \alpha_{i+1}) = p_E^{(+)}(x, \alpha_i)$. Substituting this relation in Eq. (14.20), we obtain the following recursion relation for $p_E^{(-)}(x, \alpha_i)$:

$$p_E^{(-)}(x, \alpha_i) = \frac{i\,W(x, \alpha_i)\, p_{E-R(\alpha_i)}^{(-)}(x, \alpha_{i+1}) - W^2(x, \alpha_i) + E}{p_{E-R(\alpha_i)}^{(-)}(x, \alpha_{i+1}) + i\,W(x, \alpha_i)} \ . \tag{14.21}$$

This recursion relation, along with the initial condition given by Eq. (14.11) determines all functions $p_E^{(-)}(x, \alpha_i)$.

To provide a concrete example, let us determine the quantum momentum function related to the first excited state of the system with energy $E = R(\alpha_1)$. In Eq. (14.21), let us substitute $E - R(\alpha_2) = 0$. Then

---

[4]This is due to the unbroken nature of supersymmetry which implies that there is no normalizable $\psi^{(+)}$ at zero energy.

$$p_{E-R(\alpha_2)}^{(-)} = p_0^{(-)} = iW$$

$$-i\,p_{R(\alpha_1)}^{(-)}(x,\alpha_1) = \frac{W(x,\alpha_2) \cdot W(x,\alpha_1) + W^2(x,\alpha_1) - R(\alpha_1)}{W(x,\alpha_1) + W(x,\alpha_2)} \quad,$$

$$= W(x,\alpha_1) - \frac{R(\alpha_1)}{W(x,\alpha_1) + W(x,\alpha_2)} \quad. \qquad (14.22)$$

Therefore, starting from zero-energy QMF's $p_0^{(-)}(x,\alpha_i) = iW(x,\alpha_i)$, we have derived the higher level QMF; i.e., $p_{R(\alpha_1)}^{(-)}(x,\alpha_1)$. This procedure can be iterated to generate $p_E^{(-)}(x,\alpha_1)$ for any eigenvalue E. Using the recursion relation (14.21) we can write

$$p_E^{(-)}(x,\alpha_1)\frac{i\,W(x,\alpha_1)\,p_{E-R(\alpha_1)}^{(-)}(x,\alpha_2) - W^2(x,\alpha_1) + E}{p_{E-R(\alpha_1)}^{(-)}(x,\alpha_2) + i\,W(x,\alpha_1)} \qquad (14.23)$$

$$p_{E-R(\alpha_1)}^{(-)}(x,\alpha_2) = \frac{i\,W(x,\alpha_2)\,p_{E-R(\alpha_1)-R(\alpha_2)}^{(-)}(x,\alpha_3)}{p_{E-R(\alpha_1)-R(\alpha_2)}^{(-)}(x,\alpha_3) + i\,W(x,\alpha_2)}$$
$$- \frac{W^2(x,\alpha_2) - E - R(\alpha_1)}{p_{E-R(\alpha_1)+R(\alpha_2)}^{(-)}(x,\alpha_3) + i\,W(x,\alpha_2)} \quad.$$

$$\qquad (14.24)$$

Substituting (14.24) into (14.23) we obtain

$$p_E^{(-)}(x,\alpha_1) = \frac{A_3\,p_{E-R(\alpha_2)-R(\alpha_1)}^{(-)}(x,\alpha_3) + B_3}{C_3\,p_{E-R(\alpha_2)-R(\alpha_1)}^{(-)}(x,\alpha_3) + D_3} \qquad (14.25)$$

where

$A_3 = E - R(\alpha_1) - W(x,\alpha_2)W(x,\alpha_1) - W^2(x,\alpha_2)$,

$B_3 = iW(x,\alpha_2)\left(W^2(x,\alpha_1) - E\right) + iW(x,\alpha_1)\left(W^2(x,\alpha_2) - E + R(\alpha_1)\right)$,

$C_3 = -i\,W(x,\alpha_2) - i\,W(x,\alpha_1)$,

$D_3 = E - W(x,\alpha_2)W(x,\alpha_1) - W^2(x,\alpha_1)$.

The recursion relation given by Eq. (14.25) is of the type

$$z_{i+1} = \frac{az_i + b}{cz_i + d} \quad. \qquad (14.26)$$

Such relations are called fractional linear transformation in z. The composition of two fractional linear transformations is tedious to compute. A short-cut is provided by the map

$$\frac{az + b}{cz + d} \mapsto \begin{bmatrix} a & b \\ c & d \end{bmatrix} \quad. \qquad (14.27)$$

It is easy to check that the function composition corresponds to matrix multiplication. That is, if f_1 and f_2 are two transformations given by

$$f_1(z) = \frac{a_1 z + b_1}{c_1 z + d_1}, \quad f_2(z) = \frac{a_2 z + b_2}{c_2 z + d_2}, \tag{14.28}$$

then

$$(f_2 \circ f_1)(z) = \frac{az + b}{cz + d}, \tag{14.29}$$

where the coefficients a, b, c and d are given by

$$\begin{bmatrix} a & b \\ c & d \end{bmatrix} = \begin{bmatrix} a_2 & b_2 \\ c_2 & d_2 \end{bmatrix} \cdot \begin{bmatrix} a_1 & b_1 \\ c_1 & d_1 \end{bmatrix}. \tag{14.30}$$

For any transformation f of form (14.26) there exists an inverse transformation f^{-1} if and only if $ad - bc \neq 0$. In this case the matrix associated with f is nonsingular and its inverse gives the coefficients of f^{-1}.

Now, let us associate to transformation (14.23) the matrix

$$m_1 = \begin{bmatrix} i\,W(x, \alpha_1) & E - W^2(x, \alpha_1) \\ 1 & i\,W(x, \alpha_1) \end{bmatrix}. \tag{14.31}$$

Similarly, we associate to transformation (14.24) the matrix

$$m_2 = \begin{bmatrix} i\,W(x, \alpha_2) & E - R(\alpha_1) - W^2(x, \alpha_2) \\ 1 & i\,W(x, \alpha_2) \end{bmatrix}. \tag{14.32}$$

Then the coefficients A_3, B_3, C_3, D_3 of Eq. (14.25) are obtained from

$$\begin{bmatrix} A_3 & B_3 \\ C_3 & D_3 \end{bmatrix} = m_2 \cdot m_1. \tag{14.33}$$

Using the fractional linear transformation property, we can easily generalize this result. We have

$$p_E^{(-)}(x, \alpha_1) = \frac{A_{n+1}\, p_{E - \sum_{i=1}^{n} R(\alpha_i)}^{(-)}(x, \alpha_{n+1}) + B_{n+1}}{C_{n+1}\, p_{E - \sum_{i=1}^{n} R(\alpha_i)}^{(-)}(x, \alpha_{n+1}) + D_{n+1}}, \tag{14.34}$$

where

$$\begin{bmatrix} A_{n+1} & B_{n+1} \\ C_{n+1} & D_{n+1} \end{bmatrix} = m_n \cdot m_{n-1} \cdots m_1 \tag{14.35}$$

and

$$m_k = \begin{bmatrix} i\,W(x, \alpha_k) & E - \sum_{j=1}^{k-1} R(\alpha_j) - W^2(x, \alpha_k) \\ 1 & iW(x, \alpha_k) \end{bmatrix},$$

$$k = 1, 2, \ldots, n. \tag{14.36}$$

Note that the determinant of m_k is given by

$$\det (m_k) = E - \sum_{j=1}^{k-1} R(\alpha_j) \ . \tag{14.37}$$

Therefore, for those values of energy where $E = \sum_{j=1}^{k-1} R(\alpha_j)$ the matrix m_k is singular.

We have now connected QHJ Theory with supersymmetric quantum mechanics, and have shown that the quantum momenta of supersymmetric partner potentials are connected via linear fractional transformations. Then, by making use of the matrix representation of the linear fractional transformations, we have provided an algorithm to generate any quantum momentum function for a general shape invariant potential. In the the next section we will study the constraints placed upon the singularity structure of QMF by shape invariance condition.

14.2   Determination of Eigenvalues using Shape Invariance

In the last section we showed that in QHJ, shape invariance enables us to recursively determine the quantum momenta, and hence the eigenfunctions, analogous to SUSYQM. However, this derivation assumed that the eigenvalues E_n were given by the sum $\sum_{i=1}^{n-1} R(\alpha_i)$. Let us examine how the shape invariance condition also provides sufficient information to determine the singularity structure of the quantum momentum functions, and thus helps in determining eigenvalues of the system via QHJ as well.

The shape invariance condition, when written in terms of the superpotential W, takes the form:

$$W^2(x,\alpha_0) + \frac{dW(x,\alpha_0)}{dx} = W^2(x,\alpha_1) - \frac{dW(x,\alpha_1)}{dx} + g(\alpha_1) - g(\alpha_0). \tag{14.38}$$

How does the above condition affect the pole structure of the superpotential? We will see that its behavior near fixed poles is constrained. In our analysis, we will divide all superpotentials into two categories: those with infinite domain $(0, \infty)$ or $(-\infty, \infty)$, and those with finite domain, such as (x_1, x_2). In QHJ formalism we embed these domains into a complex plane and analyze the behavior of QMF's near singular points. The superpotentials with finite domain generate infinitely many copies of the same domain on the real axis. The pole structure of QMF's is also repeated infinitely many times on the real axis. One of the ways to solve these systems is to map these finite domains to the infinite domains and thus turn them into an infinite-domain problem. For this reason, we will only consider systems with infinite domains.

Our objective in this section is to show that using shape invariance and QHJ formalism, we can derive spectra for all infinite domain systems considered in Chapter 6. We divide all superpotentials into two categories: a) those that are of algebraic form and b) those can be converted into an algebraic form using a transformation such as $y = e^x$. We will call them algebraic and exponential respectively.

14.3 Algebraic Cases

Let us first consider the superpotentials that can be explicitly expressed in an algebraic form. We would like to determine their structure near the fixed singularities, using shape invariance. In particular, we will be exploring how the superpotential diverges, if at all, near the origin and infinity. Let us first analyze the superpotential in the vicinity of the origin. Expanding in powers of the coordinate r,

$$
\begin{aligned}
W(r, \alpha_0) &= \sum_{k=-1}^{\infty} b_k(\alpha_0) \, r^k \\
&= \frac{b_{-1}(\alpha_0)}{r} + b_0(\alpha_0) \, r^0 + b_1(\alpha_0) \, r^1 + b_2(\alpha_0) \, r^2 + \cdots .
\end{aligned}
$$

We have not included a term more singular the $1/r$ in the superpotential to avoid the particle's "falling into the center."[5] The derivative of the superpotential is then given by

$$
\frac{dW(r, \alpha_0)}{dr} = -\frac{b_{-1}(\alpha_0)}{r^2} + 0 + b_1(\alpha_0) + 2\,b_2(\alpha_0)\,r + \cdots ;
$$

and the square of the superpotential is given by

$$
\begin{aligned}
W^2(r, \alpha_0) &= (b_{-1}(\alpha_0))^2 \, r^{-2} + (2\,b_{-1}(\alpha_0)\,b_0(\alpha_0))\,r^{-1} \\
&\quad + \left((b_0(\alpha_0))^2 + 2\,b_{-1}(\alpha_0)\,b_1(\alpha_0) \right) + \cdots
\end{aligned}
$$

Substituting these expansions into the shape invariance condition (14.38), we arrive at the following constraints on coefficients of the superpotential $W(x, \alpha_0)$:

$$
b_{-1}(\alpha_0)^2 - b_{-1}(\alpha_0) = b_{-1}(\alpha_1)^2 + b_{-1}(\alpha_1) , \tag{14.39}
$$

$$
2\,b_{-1}(\alpha_0)\,b_0(\alpha_0) = 2\,b_{-1}(\alpha_1)\,b_0(\alpha_1) , \tag{14.40}
$$

---

[5] See the discussion of such potentials in Chapter 12.

$$b_0(\alpha_0)^2 + 2\,b_{-1}(\alpha_0)\,b_1(\alpha_0) + b_1(\alpha_0)$$
$$= \quad b_0(\alpha_1)^2 + 2\,b_{-1}(\alpha_1)\,b_1(\alpha_1) - b_1(\alpha_1) + R(\alpha_0) \ .$$
$$(14.41)$$

The constraint (14.39) is a quadratic equation for the coefficient $b_{-1}(\alpha_1)$ in terms of $b_{-1}(\alpha_0)$. It has the following two solutions: $b_{-1}(\alpha_1) = -b_{-1}(\alpha_0)$ or $b_{-1}(\alpha_1) = b_{-1}(\alpha_0) - 1$. Since $b_{-1}(\alpha_0)$ is the coefficient of the dominant term near the origin, its behavior plays an essential role in deciding whether supersymmetry remains unbroken. If the supersymmetry is unbroken for a given value of the coefficient $b_{-1}(\alpha_0)$, changing the sign of this coefficient would change the asymptotic value of the superpotential as $r \to 0$, and supersymmetry will no longer remain unbroken. Hence, the first solution, $b_{-1}(\alpha_1) = -b_{-1}(\alpha_0)$, is not acceptable if both $b_{-1}(\alpha_0)$ and $b_{-1}(\alpha_1)$ were to keep the system in the parameter domain needed for unbroken supersymmetry. Thus, the constraint (14.39) implies the relationship $b_{-1}(\alpha_1) = b_{-1}(\alpha_0) - 1$. For $\alpha_1 = \alpha_0 + 1$, this gives the difference equation $b_{-1}(\alpha_0 + 1) = b_{-1}(\alpha_0) - 1$, whose solution is

$$b_{-1}(\alpha_0) = -\alpha_0 + \text{constant} \ . \qquad (14.42)$$

We choose the constant to be zero. Eq. (14.40) then leads to the difference equation

$$\alpha_0\,b_0(\alpha_0) = \alpha_1\,b_0(\alpha_1) \ . \qquad (14.43)$$

In other words, the product $\alpha_0\,b_0(\alpha_0)$ does not depend on the parameter α_0; we denote this product by β. Hence, we have

$$b_0(\alpha_0) = \frac{\beta}{\alpha_0} \ .$$

Thus, near the origin the structure of the superpotential $W(x, \alpha_0)$ is given by:

$$W\left(r, \alpha_0\right)\big|_{\text{near origin}} = -\frac{\alpha_0}{r} + \frac{\beta}{\alpha_0} + \cdots \ . \qquad (14.44)$$

Now, let us explore the structure of the superpotential near infinity. To do this, as we have done before, we define $u = \frac{1}{r}$. The shape invariance condition of Eq. (14.38) in terms of the variable u transforms into

$$W^2\left(\frac{1}{u}, \alpha_0\right) - u^2\,\frac{dW\left(\frac{1}{u}, \alpha_0\right)}{du} = W^2\left(\frac{1}{u}, \alpha_1\right) + u^2\,\frac{dW\left(\frac{1}{u}, \alpha_1\right)}{du} + R(\alpha_0) \ .$$
$$(14.45)$$

Problem 14.1. *Derive Eq. (14.45).*

Analogous to the expansion near the origin, let us expand $W\left(\frac{1}{u}, \alpha_0\right)$ in powers of u as:

$$W\left(1/u, \alpha_0\right) = \sum_{k=-1}^{\infty} c_k(\alpha_0)\, u^k$$

$$= \frac{c_{-1}(\alpha_0)}{u} + c_0(\alpha_0)\, u^0 + c_1(\alpha_0)\, u^1 + c_2(\alpha_0)\, u^2 + \cdots.$$

Similarly expanding W^2 and $\frac{dW(r,\alpha_0)}{dr}$, we obtain

$$W^2\left(1/u, \alpha_0\right) = \frac{(c_{-1}(\alpha_0))^2}{u^2} + \frac{(2\, c_{-1}(\alpha_0)\, c_0(\alpha_0))}{u}$$

$$+ \left((c_0(\alpha_0))^2 + 2\, c_{-1}(\alpha_0)\, c_1(\alpha_0)\right) + \cdots$$

and

$$\frac{dW(r, \alpha_0)}{dr} = -u^2\, \frac{dW(1/u, \alpha_0)}{du}$$

$$= c_{-1}(\alpha_0) - c_1(\alpha_0)\, u^2 - 2\, c_2(\alpha_0)\, u^3 + \cdots.$$

Now we substitute these expressions into the transformed shape invariance condition of Eq. (14.45) to get

$$c_{-1}(\alpha_0)^2 = c_{-1}(\alpha_1)^2 \tag{14.46}$$

$$2\, c_{-1}(\alpha_0)\, c_0(\alpha_0) = 2\, c_{-1}(\alpha_1)\, c_0(\alpha_1) \tag{14.47}$$

$$c_0(\alpha_0)^2 + 2\, c_{-1}(\alpha_0)\, c_1(\alpha_0) + c_{-1}(\alpha_0)$$

$$= c_0(\alpha_1)^2 + 2\, c_{-1}(\alpha_1)\, c_1(\alpha_1) - c_{-1}(\alpha_1) + R(\alpha_0). \tag{14.48}$$

Problem 14.2. *Derive Eqs. (14.46)–(14.48).*

From Eq. (14.46), we obtain $c_{-1}(\alpha_1) = \pm c_{-1}(\alpha_0)$. As we argued before, since we assumed that supersymmetry remains unbroken for a range of values of the parameter, we choose $c_{-1}(\alpha_1) = c_{-1}(\alpha_0)$; i.e., this coefficient does not depend on the parameter α_0. We denote it by c_{-1}. This then leads to $c_0(\alpha_0) = c_0(\alpha_1) = c_0$, another constant. Eq. (14.48) now yields $2\, c_{-1}\, c_1(\alpha_0) + c_{-1} = 2\, c_{-1}\, c_1(\alpha_1) - c_{-1} + g(\alpha_1) - g(\alpha_0)$, which gives

$$c_1(\alpha_1) + \frac{g(\alpha_1)}{2\, c_{-1}} = c_1(\alpha_0) + \frac{g(\alpha_0)}{2\, c_{-1}} + 1. \tag{14.49}$$

This difference equation is of the form

$$f(\alpha_1) = f(\alpha_0) + 1 , \qquad (14.50)$$

where $f(\alpha_0) = c_1(\alpha_0) + \frac{g(\alpha_0)}{2\,c_{-1}}$. Since $\alpha_1 = \alpha_0 + 1$, the solution of Eq. (14.50) is $f(\alpha_0) = \alpha_0 + \Delta$. Hence, we have $c_1(\alpha_0) = \alpha_0 - \frac{g(\alpha_0)}{2\,c_{-1}} + \Delta$, where Δ is a constant independent of α_0. Thus, near infinity the superpotential is given by

$$\widetilde{W}\left(\frac{1}{u}, \alpha_0\right)\Big|_{\text{infinity}} = \frac{c_{-1}}{u} + c_0$$

$$+ \left(\alpha_0 - \frac{g(\alpha_0)}{2\,c_{-1}} + \Delta\right) u + \ldots . \qquad (14.51)$$

The result we have obtained depended crucially upon the assumption that c_{-1} is not zero. If $c_{-1} = 0$, the structure of the potential near infinity will be very different and the constraints given by Eqs. (14.46 - 14.48) would no longer be valid. We present this case next.

For $c_{-1} = 0$, the superpotential near infinity is given by $\widetilde{W}(u, \alpha_0) = c_0 + c_1 u + c_2 u^2 + \cdots$. This leads to $\widetilde{W}^2(u, \alpha_0) = c_0^2 + 2c_0 c_1 u + (c_1^2 + 2c_0 c_2)u^2 + \cdots$, and $u^2 \frac{d\widetilde{W}}{du} = c_1 u^2 + 2c_2 u^3 + \cdots$. Since $\widetilde{W}^2(u, \alpha_0)$ and $\widetilde{W}^2(u, \alpha_1)$ must satisfy the shape invariance condition (14.38), matching the first two powers of u we obtain

$$c_0^2(\alpha_1) = c_0^2(\alpha_1) + R(\alpha_0) , \qquad (14.52)$$

$$c_0(\alpha_0)c_1(\alpha_0) = c_0(\alpha_1)c_1(\alpha_1) . \qquad (14.53)$$

Eq. (14.52) may be rewritten as $c_0^2(\alpha_0) + g(\alpha_0) = c_0^2(\alpha_0+1) + g(\alpha_0+1)$ which means that the quantity $c_0^2(\alpha_0) + g(\alpha_0)$ is independent of the argument α_0, and hence equal to a constant Λ. This leads to

$$c_0(\alpha_0) = \pm\sqrt{-g(\alpha_0) + \Lambda} . \qquad (14.54)$$

Now from Eq. (14.53) we see that the product $c_0(\alpha_0)c_1(\alpha_0)$ is also independent of the argument α_0, and hence it must also be equal to a constant, which we denote by β. From $c_0(\alpha_0)c_1(\alpha_0) = \beta$, we have

$$c_1(\alpha_0) = \frac{\beta}{c_0(\alpha_0)} = \frac{\beta}{\pm\sqrt{-g(\alpha_0) + \Lambda}} . \qquad (14.55)$$

Thus, near ∞,

$$\widetilde{W}(u, \alpha_0) = \pm\left(\sqrt{-g(\alpha_0) + \Lambda}\right) \pm \left(\frac{\beta}{\sqrt{-g(\alpha_0) + \Lambda}}\right) u + \cdots . \qquad (14.56)$$

We have now acquired significant knowledge about the structure of the superpotential near boundary points simply from the requirement of shape invariance. Let us write these findings again. Near the origin, as given in Eq. (14.44), the structure is

$$W(r, \alpha_0) = -\frac{\alpha_0}{r} + \frac{\beta}{\alpha_0} + \cdots .\tag{14.57}$$

Near infinity, as $u \to 0$, there are two possible structures that depend on whether the superpotential has a singularity; i.e., whether the value of the coefficient $c_{-1}(\alpha_0)$ is zero or non-zero. They are:

$$\widetilde{W}(u, \alpha_0) = \begin{cases} \frac{c_{-1}}{u} + c_0 + \left(\alpha_0 - \frac{g(\alpha_0)}{2\,c_{-1}} + D\right) u + \cdots & c_{-1}(\alpha_0) \neq 0 \\[2mm] \pm\sqrt{-g(\alpha_0) + \Lambda} \pm \frac{\beta}{\sqrt{-g(\alpha_0)+\Lambda}}\, u + \cdots & c_{-1}(\alpha_0) = 0. \end{cases}$$
$$\tag{14.58}$$

With this understanding about the structure of $W(x, \alpha_0)$ at the end points of the domain, we now proceed to substitute Eqs. (14.57) and (14.58) into QHJ Eq. (14.6). This substitution will help us learn about their implications for the QMF, and thence the eigenvalues of the hamiltonian. We first expand the momentum $p(r)$ near the origin as:

$$p(r, \alpha_0) = \sum_{k=-1}^{\infty} p_k\, r^k = \frac{p_{-1}}{r} + p_0\, r^0 + p_1\, r^1 + p_2\, r^2 + \cdots .$$

The derivative of p is then given by $\frac{d\,p\,(r,\alpha_0)}{dr} = -\frac{p_{-1}}{r^2} + 0 + p_1 + 2\,p_2\, r + \cdots$ and the square of p is given by $p^2(r, \alpha_0) = \frac{(p_{-1})^2}{r^2} + \frac{(2\,p_{-1}\,p_0)}{r} + (p_0^2 + 2\,p_{-1}\,p_1) + \cdots$. Collecting terms with various powers of r, we find that near the origin the combination $p^2 + ip'$ is given by

$$\left(p^2 - ip'\right)\big|_{\text{origin}} \sim \frac{p_{-1}^2 + ip_{-1}}{r^2} + \frac{(2\,p_{-1}\,p_0)}{r} + \left(p_0^2 + 2\,p_{-1}\,p_1 + ip_1\right) + \cdots .$$

We substitute this series in the QHJ Eq. (14.10) and equate the coefficients of powers of r on both sides. From the coefficient of r^{-2}, we obtain get $p_{-1}^2 + ip_{-1} + \alpha_0^2 - \alpha_0 = 0$. It has two solutions:

$$p_{-1} = i(\alpha_0 - 1) \text{ and } p_{-1} = i\alpha_0 .\tag{14.59}$$

Using Eq. (14.11), we choose $p_{-1} = -i\alpha_0$. We will later see that only this term in the expansion of p will contribute when we carry out the contour integration around the origin in the complex r plane. So we do not list other coefficients.

To determine the leading order behavior of the momentum function at infinity, let us first consider the case that $c_1(\alpha_0) \neq 0$. We expand the momentum function $p\left(\frac{1}{u}, \alpha_0\right) \equiv \widetilde{p}(u, \alpha_0)$ in powers of u:

$$\widetilde{p}(u, \alpha_0) = \sum_{k=-1}^{\infty} \widetilde{p}_k u^k = \frac{\widetilde{p}_{-1}}{u} + \widetilde{p}_0 u^0 + \widetilde{p}_1 u^1 + \widetilde{p}_2 u^2 + \cdots .$$

This leads to the following expression for the combination $p^2 - ip'$:

$$(p^2 - ip')\big|_\infty \equiv \widetilde{p}^2 + iu^2 \frac{d\widetilde{p}}{du} = \frac{\widetilde{p}_{-1}^2}{u^2} + \frac{2\widetilde{p}_{-1}\widetilde{p}_0}{u}$$
$$+ \left(\widetilde{p}_0^2 + 2\widetilde{p}_{-1}\widetilde{p}_1 - i\widetilde{p}_{-1}\right) + \cdots .$$

$$(14.60)$$

Substituting it into Eq. (14.10); i.e., in

$$\widetilde{p}^2 + iu^2 \frac{d\widetilde{p}}{du} = E - \left(\widetilde{W}^2 + u^2 \frac{d\widetilde{W}}{du}\right) ,$$

$$(14.61)$$

we obtain

$$\frac{\widetilde{p}_{-1}^2}{u^2} + \frac{2\widetilde{p}_0\widetilde{p}_{-1}}{u} + \left(\widetilde{p}_0^2 + 2\widetilde{p}_{-1}\widetilde{p}_1 - i\widetilde{p}_{-1}\right) + \cdots$$
$$= -\frac{c_{-1}^2}{u^2} - \frac{2c_0 c_{-1}}{u} + E - c_0^2 - 2c_{-1}\left(\alpha_0 - \frac{g(\alpha_0)}{2c_{-1}}\right) + c_{-1} .$$

By identifying the various powers of the variable u, we obtain $\widetilde{p}_{-1} = ic_{-1}$; $\widetilde{p}_0 = ic_0$; and $\widetilde{p}_1 = i\alpha_0 - i\frac{g(\alpha_0)+E}{2c_{-1}}$. At infinity, as we shall soon see, it is this last coefficient $\widetilde{p}_1$ that contributes towards the contour integration.

We are now ready to determine the eigenvalue for the system using the information we have generated solely from shape invariance. The contribution to the contour integration is given by

$$\left[\oint p\,dr\right]_{\text{around moving poles}} = -\oint_{r=0} p\,dr - \oint_{r=\infty} p\,dr$$

$$= -\oint_{r=0} p\,dr + \oint_{u=0} \frac{1}{u^2}\widetilde{p}\,du$$

$$= -(2\pi i p_{-1}) + (2\pi i \widetilde{p}_1)$$

$$= -2\pi i\,(-i\,\alpha_0) + 2\pi i\left[i\,\alpha_0 - i\frac{g(\alpha_0)+E}{2c_{-1}}\right]$$

$$= 2\pi\left[-2\alpha_0 + \frac{g(\alpha_0)+E}{2c_{-1}}\right] .$$

$$(14.62)$$

Note that due to the $\frac{1}{u^2}$ factor in the term $\oint_{u=0}(\frac{1}{u^2}\,\tilde{p})\,du$, only the linear term of $\tilde{p}$, $2\pi\,i\tilde{p}_1$, contributes towards the integral.

It is worth noting that we cannot get the entire global behavior of the potential from expansions at the origin and infinity. If the potential is a symmetric function of r, as is the case for e.g. the three dimensional harmonic oscillator, we obtain an additional contribution. If we embed the domain $(0,r)$ in a two-dimensional complex plane, a mirror image of the potential function in the region $\mathcal{R}e\,(r) < 0$ would then have singularities for $p(r)$, precisely at the same points as $p(r)$ does in the region $\mathcal{R}e\,(r) < 0$. As a consequence,

$$\oint p\,dr\bigg|_{\text{Moving poles}} \equiv \oint p\,dr\bigg|_{\mathcal{R}e\,(r)>0}$$

$$= -\oint p\,dr\bigg|_{\mathcal{R}e\,(r)<0} - \oint_{r=0} p\,dr + \oint_{u=0} \frac{1}{u^2}\,\tilde{p}\,du\;.$$

However, $\oint p\,dr\big|_{\mathcal{R}e\,(r)>0}$ and $\oint p\,dr\big|_{\mathcal{R}e\,(r)<0}$ have identical value due to the symmetry of the potential. This leads to

$$\oint p\,dr\bigg|_{\text{around moving poles}} = \frac{1}{2}\left(-\oint_{r=0} p\,dr + \oint_{u=0} \frac{1}{u^2}\,\tilde{p}\,du\right) = 2\pi n\;.$$

$$(14.63)$$

From Eq. (14.62), we then get

$$\frac{1}{2}\left\{2\pi\left[-2\alpha_0 + \frac{g(\alpha_0) + E}{2\,c_{-1}}\right]\right\} = 2\pi n\;. \tag{14.64}$$

Solving for the energy, Eq. (14.62) gives

$$E_n = 4\,c_{-1}\,(n + \alpha_0) - g(\alpha_0)\;. \tag{14.65}$$

To determine the value of $g(\alpha_0)$, we demand as per SUSYQM that the ground state energy E_0 be zero. Substituting $n = 0$, we obtain $g(\alpha_0) = 4\,c_{-1}\alpha_0$. Since $\alpha_n = \alpha_0 + n$, we obtain $g(\alpha_n) = 4\,c_{-1}\alpha_n = 4\,c_{-1}\,(\alpha_0 + n)$. Substituting for $g(\alpha_0)$ and $g(\alpha_n)$ in E_n, we obtain

$$E_n = 4n\,c_{-1} = g(\alpha_n) - g(\alpha_0). \tag{14.66}$$

Thus, QHJ formalism also generates the correct eigenvalues from shape invariance. For the three-dimensional harmonic oscillator, we identify the coefficient c_{-1} with $\frac{1}{2}\,\omega$. As expected, the energy is

$$E_n = 4n\,c_{-1} = 2n\omega\;. \tag{14.67}$$

Problem 14.3. *In the above derivation, we assumed that the origin was a singular point. Repeat the procedure without a singular point at the origin. Show that for a shape invariant potential with only one fixed singularity, at infinity, the eigenvalues are same as that of a one-dimensional harmonic oscillator; i.e., $E_n \propto n$.*

Now let us consider the case in which the superpotential has no divergence at infinity; i.e., $c_{-1} = 0$. For this case, we will need to determine the structure of the quantum momentum function $p(x)$ at infinity. (The structure at the origin remains the same as that for $c_{-1} \neq 0$.) We substitute the expansion of the superpotential $\widetilde{W}(u, \alpha_0)$ from Eq. (14.58) into the QHJ equation at infinity given by Eq. (14.61), and we obtain

$$\frac{\tilde{p}_{-1}^2}{u^2} + \frac{2\tilde{p}_0 \tilde{p}_{-1}}{u} + (\tilde{p}_0^2 + 2\tilde{p}_{-1}\tilde{p}_1 - i\tilde{p}_{-1}) + \cdots = E + g(\alpha_0) - \Lambda - 2\beta u + \cdots .$$

Now equating the coefficients of powers of u, we obtain $\tilde{p}_{-1}(\alpha_0) = 0$; $\tilde{p}_0^2(\alpha_0) = E + g(\alpha_0) - \Lambda$; $\tilde{p}_0(\alpha_0)\tilde{p}_1(\alpha_0) = \beta$; etc. From these, we obtain the following coefficients:

$$\tilde{p}_0 = \pm\sqrt{E + g(\alpha_0) - \Lambda}; \quad \text{and} \quad \tilde{p}_1 = \pm\frac{\beta}{\sqrt{E + g(\alpha_0) - \Lambda}} .$$

The quantization condition then yields

$$\oint p\, dr\Big|_{\text{Moving poles}} = -\oint_{r=0} p\, dr + \oint_{u=0} \frac{1}{u^2}\, \tilde{p}\, du \tag{14.68}$$

$$= -(2\pi i p_{-1}) + (2\pi i \tilde{p}_1)$$

$$= -2\pi i\,(-i\alpha_0) \pm 2\pi i\left(\frac{\beta}{\sqrt{E + g(\alpha_0) - \Lambda}}\right)$$

$$= 2\pi n .$$

Thus, we have $-\alpha_0 \pm i\dfrac{\beta}{\sqrt{E+g(\alpha_0)-\Lambda}} = n$. Solving for E_n,

$$E_n = \Lambda - g(\alpha_0) - \frac{\beta}{(n + \alpha_0)^2} . \tag{14.69}$$

Since we are in the domain of unbroken supersymmetry, we must have $E_0 = 0$. Hence substituting $n = 0$ in Eq. (14.69), we obtain $g(\alpha_0) = -\frac{1}{\alpha_0^2} + \Lambda$; i.e., $g(\alpha_n) = -\frac{1}{(\alpha_0+n)^2} + \Lambda$. The energy E_n is

$$E_n = \beta^2\left[\frac{1}{(\alpha_0)^2} - \frac{1}{(n + \alpha_0)^2}\right] = g(\alpha_n) - g(\alpha_0). \tag{14.70}$$

Thus, we have determined the energy of this system described by a Coulomb-like potential that is singular at the origin and has no singularity at infinity, simply by using the shape invariance condition. Once we identify β^2 with e^2 and α_0 with $l + 1$, we obtain the energy spectrum for the Coulomb potential.

14.4   Exponential Cases

Here we consider superpotentials that arise as a linear combination of various powers of the exponential function e^x. For example, the Morse and Rosen-Morse potentials fall into this category. We assume that our generic superpotential has the general form

$$W(x, \alpha_0) = \sum_j b_j \left[e^x\right]^j. \qquad (14.71)$$

This superpotential has singularities at $\pm\infty$. To analyze the nature of these singularities we first convert it into an algebraic form with the transformation $r = e^x$. This maps $x \in (-\infty, \infty)$ to $r \in (0, \infty)$. We then subject the resulting superpotential to the shape invariance condition. As $x \to -\infty$, and thus $r \to 0$, we assume the potential has the following structure:

$$W(r, \alpha_0) = b_{-1}r^{-1} + b_0 + b_1 r + \cdots. \qquad (14.72)$$

Similarly, as $r \to \infty$, we express the superpotential in terms of a variable $u \equiv 1/r$:

$$\widetilde{W}(u, \alpha_0) = c_{-1}u^{-1} + c_0 + c_1 u + \cdots. \qquad (14.73)$$

Now we will determine how shape invariance constrains various terms in these two expansions at the end points of the domain of r.

As with the algebraic superpotential, the following procedure differs depending on whether b_{-1} and c_{-1} are each zero or nonzero, making for four possible cases. We want to stress that we are considering only those shape invariant superpotentials that have at least one singularity at either the origin or infinity. *If the superpotential has no poles at these extreme points, then it must have poles at "finite" fixed-points.* This type of potential would fall under what we call trigonometric potentials and will not be considered.

It is interesting to note that the case $b_{-1} = 0$, and $c_{-1} \neq 0$ is identical to the case $b_{-1} \neq 0$, and $c_{-1} = 0$. Hence, we need consider only two cases [6]: $(b_{-1} \neq 0, \ c_{-1} = 0)$ and $(b_{-1} \neq 0, \ c_{-1} \neq 0)$.

Problem 14.4. *Show that the above claim is true.*

[6]The case $(b_{-1} = 0, \ c_{-1} = 0)$ has no fixed pole on the real-axis.

However, it turns out that for $(b_{-1} \neq 0,\ c_{-1} \neq 0)$, the quantization condition $\int p\, dx = 2\pi n$ does not yield an equation involving energy, and hence energy cannot be determined. There are no known exponential-type shape invariant potentials in quantum mechanics for which the superpotential has singularities for both positive and negative infinity. Thus, we will only consider the case where $b_{-1} \neq 0$ and $c_{-1} = 0$.

Near $r = 0$, the superpotential is given by

$$W(r, \alpha_0) = \frac{b_{-1}(\alpha_0)}{r} + b_0(\alpha_0) + b_1(\alpha_0)r + \cdots . \tag{14.74}$$

From W, we obtain

$$\frac{dW}{dx} = \frac{dW}{dr}\frac{dr}{dx} = r\frac{dW}{dr} = -\frac{b_{-1}}{r} + b_1 r , \tag{14.75}$$

and

$$W^2 = \frac{b_{-1}^2}{r^2} + \frac{2b_{-1}b_0}{r} + b_0^2 + 2b_{-1}b_1 + \cdots . \tag{14.76}$$

Hence, due to the above change of variable, the shape invariance condition now takes the form

$$W^2(\alpha_0) + r\frac{dW(\alpha_0)}{dr} = W^2(\alpha_1) - r\frac{dW(\alpha_1)}{dr} + R(\alpha_0). \tag{14.77}$$

Substituting the expansion of W into this equation and equating powers of r, we obtain

$$b_{-1}^2(\alpha_0) = b_{-1}^2(\alpha_1) ,$$

$$2b_{-1}(\alpha_0)b_0(\alpha_0) - b_{-1}(\alpha_0) = 2b_{-1}(\alpha_1)b_0(\alpha_1) + b_{-1}(\alpha_1) , \tag{14.78}$$

$$b_0^2(\alpha_0) + 2b_{-1}(\alpha_0)b_1(\alpha_0) = b_0^2(\alpha_1) + 2b_{-1}(\alpha_1)b_1(\alpha_1) + R(\alpha_0) .$$

From the first equation, we see that $b_{-1}(\alpha_0)$ does not depend upon its argument; i.e., it is a constant. We denote this constant by b_{-1}. The second equation then gives $b_0(\alpha_1) = b_0(\alpha_0) - 1$. Substituting $\alpha_1 = \alpha_0 + 1$, we obtain the difference equation

$$b_0(\alpha_0 + 1) = b_0(\alpha_0) - 1,$$

whose solution is $b_0(\alpha_0) = -\alpha_0 + C$, where C is a constant. Now, from the last equation of the set we have

$$(-\alpha_0 + C)^2 + 2b_{-1}b_1(\alpha_0) = (-\alpha_1 + C)^2 + 2b_{-1}b_1(\alpha_1) + g(\alpha_1) - g(\alpha_0).$$

This equation can be written as

$$b_1(\alpha_0) + \frac{g(\alpha_0) + (-\alpha_0 + C)^2}{2b_{-1}} = b_1(\alpha_1) + \frac{g(\alpha_1) + (-\alpha_1 + C)^2}{2b_{-1}},$$

which shows that the expression $b_1(\alpha_0) + \frac{g(\alpha_0)+(-\alpha_0+C)^2}{2b_{-1}}$ is independent of its argument α_0, hence we set it equal to a constant D. Thus,

$$b_1(\alpha_0) = D - \frac{g(\alpha_0) + (-\alpha_0 + C)^2}{2b_{-1}} . \tag{14.79}$$

We carry out a very similar analysis in the region $r \to \infty$, where the superpotential has the form $\widetilde{W}(u, \alpha_0) = c_0(\alpha_0) + c_1(\alpha_0)\, u + c_1(\alpha_0)\, u^2 \cdots$, and we obtain

$$c_0(\alpha_0)^2 = c_0(\alpha_1)^2 + R(\alpha_0) . \tag{14.80}$$

Writing this equation as $c_0(\alpha_0)^2 + g(\alpha_0) = c_0(\alpha_1)^2 + g(\alpha_1)$, we see that the expression $c_0(\alpha_0)^2 + g(\alpha_0)$ is independent of (α_0). We set it equal to a constant Θ. Thus,

$$c_0(\alpha_0) = \pm\sqrt{\Theta - g(\alpha_0)} . \tag{14.81}$$

Therefore, asymptotic forms of the superpotential are:
Near the origin

$$W(r, \alpha_0) = \frac{b_{-1}}{r} + (-\alpha_0 + C) + \left[D - \left(\frac{g(\alpha_0) + (-\alpha_0 + C)^2}{2b_{-1}}\right)\right] r + \cdots \ ;$$
$$\tag{14.82}$$

and as $r \to \infty$ $(u \to 0)$

$$\widetilde{W}(u, \alpha_0) = \pm\sqrt{\Theta - g(\alpha_0)} + c_1(\alpha_0)\, u + c_1(\alpha_0)\, u^2 \cdots . \tag{14.83}$$

We will now substitute these two asymptotic forms of the superpotential into the QHJ equation to determine the analytic structure of p and from it the eigenvalues of the shape invariant system.

Near $r = 0$, the QHJ equation is given by

$$p^2 - i\, r\frac{dp}{dr} = E - W^2(r, \alpha_0) + r\frac{dW(r, \alpha_0)}{dr}, \tag{14.84}$$

where the superpotential $W(r, \alpha_0)$ is given by Eq. (14.82). Expanding the quantum momentum function as

$$p(r) = p_{-1}r^{-1} + p_0 + p_1 r + \cdots , \tag{14.85}$$

and substituting it in the above QHJ equation, we obtain

$$p_{-1}^2 = -b_{-1}^2; \quad 2p_{-1}p_0 + i\, p_{-1} = -b_{-1} - 2b_{-1}b_0(\alpha_0); \quad \text{etc.} \tag{14.86}$$

From the first equality in above equation, we obtain $p_{-1} = \pm i b_{-1}$. However, in the limit $E \to 0$, we must have $p \to iW$. This implies $p_{-1} = i b_{-1}$. Substituting this value of p_{-1} in the second equality, we obtain $p_0 = i\, b_0(\alpha_0) = i\,(-\alpha_0 + C)$. As we will soon see, we need go no further. The coefficient p_0 is all we need to determine the contour integral around $r = 0$.

Near $r \to \infty$, we expand the QMF as $\widetilde{p}\,(u) = \widetilde{p}_{-1} u^{-1} + \widetilde{p}_0 + \widetilde{p}_1 u + \cdots$. We substitute this expansion for $\widetilde{p}(u)$ and the superpotential given by Eq. (14.83) into the QHJ equation

$$\widetilde{p}^2 + i\,u \frac{d\widetilde{p}}{du} = E - \left(\widetilde{W}^2(u, \alpha_0) + u \frac{d\widetilde{W}(u, \alpha_0)}{du} \right). \qquad (14.87)$$

Now comparing various powers of u, we obtain $\widetilde{p}_{-1} = 0$, $(\widetilde{p}_0)^2 = \left(\pm \sqrt{\Theta - g(\alpha_0)} \right)^2 - E$, etc. Thus, we obtain $\widetilde{p}_0 = \pm \sqrt{\Theta - g(\alpha_0) - E}$.

The quantization condition for this system then yields

$$\oint_{x=-\infty} p\, dx + \oint_{x=\infty} p\, dx = -2\pi n \ . \qquad (14.88)$$

Since near $x = -\infty$, the variable $r = e^x \to 0$ and as $x \to \infty$, the variable $u \equiv 1/r = e^{-x} \to 0$, we write the above condition as

$$\oint_{r=0} \left(\frac{p_0}{r} \right) dr - \oint_{u=0} \left(\frac{\widetilde{p}_0}{u} \right) du = -2\pi n \ .$$

This then gives

$$(C - \alpha_0) \pm \sqrt{\Theta - g(\alpha_0) - E_n} = n \ .$$

Solving for E_n we obtain

$$E_n = \Theta - (C - \alpha_0 - n)^2 - g(\alpha_0) \ . \qquad (14.89)$$

Since we assume that supersymmetry is unbroken, we must have $E_0 = 0$. This implies $g(\alpha_0) = \Theta - (C - \alpha_0)^2$, and hence $g(\alpha_n) = \Theta - (C - (\alpha_0 + n))^2 = \Theta - (C - \alpha_0 - n)^2$. Thus, we have

$$E_n = (C - \alpha_0)^2 - (C - \alpha_0 - n)^2 = g(\alpha_n) - g(\alpha_0) \ . \qquad (14.90)$$

If we identify $C - \alpha_0$ with the parameter A of the Morse superpotential $W = A - Be^{-x}$, we obtain the energy $E_n = A^2 - (A - n)^2$.

Thus, through several examples we have shown that in the QHJ formalism, shape invariance and unbroken SUSY completely determine the spectrum of hamiltonians. The eigenvalues are given by

$$E_n = g(\alpha_n) - g(\alpha_0) \ ,$$

exactly the expression we have seen in Chapter 6. It is true that these spectra could be gotten much more easily using SUSYQM. However, this formalism is analytic as opposed to the algebraic method we saw in Chapter 6. This QHJ quantization, which looks very similar to the JWKB semi-classical quantization procedure we saw in Chapter 13, provides us with insight into the analytical structure of the QMF's.

Chapter 15

Dirac Theory and SUSYQM

In Chapter 1, we gave a very brief description of the Dirac equation and its solutions for a point particle such as an electron in some potential, and we claimed that the equation was "intrinsically supersymmetric." Let us now give a more detailed explanation. We shall focus especially on the hydrogen atom. We choose the convention that $\hbar$ and c are equal to 1, but retain m.

15.1 Introduction

We start with a review of the "old quantum theory"; that is, the Bohr hydrogen atom. By quantizing the angular momentum: $L = n, n = 1, 2, \cdots$, Bohr was able to obtain energies $E_n = E_0/n^2 = -13.6eV/n^2$ whose differences $E_n - E_{n'}$ matched the experimental spectra quite well. The Bohr model assumes circular orbits , for which $L = mvr$ and for the Coulomb force, $E = -mv^2/2 = -L^2/2mr^2$. Thus L_n immediately determines E_n.

Experimental advances led to the discovery of extra lines in the hydrogen spectrum; in particular, doublets: two closely spaced lines where the Bohr theory predicted only one. The splitting was very small, about 4.5×10^{-5} eV as compared to the Bohr model energy differences on the order of eV. Thus, these splittings came to be called the fine structure.

Arnold Sommerfeld proposed that the fine structure could be explained by relativistic corrections to the Bohr model. For circular orbits, even the speed of the innermost electron, $v_1 \sim 2 \times 10^6 m/s$ is much smaller than the speed of light. Sommerfeld's theory allowed for non-circular orbits; in particular, elliptical orbits of various eccentricities but the same major axes, equal to the radius of Bohr's circles. Classical mechanics predicts that these will all have the same energies as the circular orbit. But Sommerfeld added in some relativistic effects, because conservation of angular momentum predicts that electrons in eccentric orbits do have high speeds close to the nucleus; the more eccentric, the higher the speed. (This is nothing but Kepler's second law.) For highly eccentric orbits, the speeds are a significant fraction of the speed of light.

Sommerfeld labeled his orbits with a quantum number $k = 1, 2, \ldots$, and obtained an expression for the fine-structure energy shift E_{nk} from the Bohr energy E_n:

$$E_{nk} = E_n \left[\frac{\alpha^2}{n} \left(\frac{1}{k} - \frac{3}{4n} \right) \right]$$

where α is the fine structure constant [1]. Thus, the doublet splittings reflected the difference in values of k for the same n [2].

Further explication of the fine structure was provided by the identification $k \equiv \ell + 1$ where ℓ is the angular momentum quantum number. In the traditional old quantum theory vector model, $L = \sqrt{\ell(\ell + 1)}$. In that model, the fine structure was accounted for by the difference in energy between states with $j = \ell + 1/2$ and those with $j = \ell - 1/2$, where j was a mysterious new quantum number. It remained for George Uhlenbeck and Samuel Goudsmit to identify the factor $1/2$ with yet another quantum number s called "spin," representing a non-classical angular momentum intrinsic to the electron. We shall discuss this in detail in the next section.

Moving on to the "new quantum mechanics," let us consider the non-relativistic model: the time-dependent Schrödinger equation

$$H\Psi(\mathbf{r}, t) \equiv -\nabla^2 \Psi(\mathbf{r}, t) + V(\mathbf{r}, t) \, \Psi(\mathbf{r}, t) = i\partial\Psi(\mathbf{r}, t)/\partial t \qquad (15.1)$$

with H a function of both $\mathbf{r}$ and t. With a time-independent potential; $V = V(r)$, we obtain the time-independent Schrödinger equation:

$$H\psi(\mathbf{r}) \equiv -\nabla^2 \psi(\mathbf{r}) + V(\mathbf{r}) \, \psi(\mathbf{r}) = E\psi(\mathbf{r}). \qquad (15.2)$$

It cannot account for spin; thus, this is an add-on to the Schrödinger wave function, written here in spherical coordinates:

$$\psi(r, \theta, \phi, s) = R_{n\ell}(r) Y_\ell^m(\theta, \phi) \chi(s). \qquad (15.3)$$

As we saw in Chapter 2, the angular functions for any central potential are the spherical harmonics. For the Coulomb potential, $V = -e^2/r$, the radial functions are the associated Laguerre polynomials.

---

[1] In standard notation $\alpha = e^2/\hbar c \simeq 1/137.036$ In our convention, $\alpha = e^2$.

[2] Apparently due to a fortuitous derivation error, Sommerfeld's expression turned out to match Dirac's fully relativistic expression more than it should have. Ya. I. Granovskii, "Sommerfeld formula and Dirac's theory", *UFN* **174**, 577 (2004).

$\chi(s)$ plays no role in the Schrödinger equation's energy eigenvalues. Its only contribution vis-a-vis energy is to add on, *ad hoc*, a contribution to the fine structure. About half of the fine structure could be accounted for by the first-order relativistic correction to the the classical kinetic energy. The rest could be accounted for by treating the "spinning" electron as a magnetic dipole in the field created by the proton orbiting the electron, as seen by the electron. Since the proton's angular momentum in the electron's frame is proportional to the orbital angular momentum **L** of the electron in the proton's frame, the so-called spin-orbit coupling is a quantity proportional to **L** · **S**.

The final result is

$$E_{fs}/E_n = \frac{\alpha^2}{n^4} \left[\frac{n}{j+1/2} - \frac{3}{4} \right] .$$

Sommerfelds' k is now identified as $j + 1/2$ rather than $\ell + 1$. Note that the fine structure energy depends only on the quantum numbers j and n. Instead of calling it E_{fs} we could just as well call it E_{nj}, and we will. In other words, it is degenerate with respect to the quantum numbers ℓ and s. Figure 15.1 shows the first few levels for hydrogen. At the bottom are the values of ℓ. Above each are the Bohr energy levels with the fine structure corrections (vastly magnified). Those levels of different ℓ but the same j have the same energy. For example, the level $\ell = 2, j = \ell - 1/2 = 3/2$ has the same energy as $\ell = 1, j = \ell + 1/2 = 3/2$.

As is evident, the diagram resembles the SUSYQM energy diagram for a series of partner potentials. We shall account for this by going beyond the approximate relativistic corrections to the Schrödinger equation, and studying the Dirac equation. It will completely account for the fine structure splittings. Later, we will see how these are the energy levels for SUSYQM partner potentials.

15.2 Spin

As we noted above, the appearance of unexpected spectral lines, the fine structure, was accounted for by postulating an intrinsic angular momentum called spin. Its magnitude was $1/2$, and it had only two possible orientations, designated "up" and "down". It resided in its own two-dimensional space, independent of the ordinary three dimensional space in which the orbital angular momentum resides.

The confirmation of the existence of this purely quantum mechanical

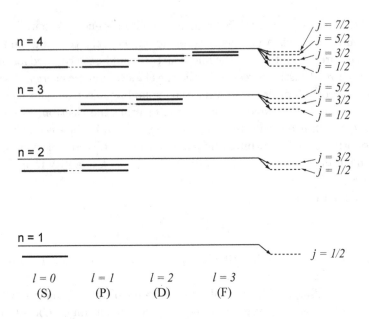

Fig. 15.1   The first few level for the hydrogen spectrum.

angular momentum was provided by Stern and Gerlach in 1922. They sent a beam of silver atoms through an inhomogeneous magnetic field, as shown in Figure 15.2 and captured them on a detector perpendicular to the initial atomic beam.

Silver has 47 electrons. The inner 46 electrons can be visualized as forming an approximately spherically symmetric closed shell, which produces no net orbital angular momentum [3]. Consequently, the magnetic moment of the silver atom is given by the magnetic moment of the 47th unpaired outer electron. This magnetic moment is proportional to the intrinsic angular momentum of the electron.

Since the potential energy of a magnetic dipole μ with a magnetic field **B** is $V = -\mu \cdot \mathbf{B}$, the force on the dipole is $-\nabla \mu \cdot \mathbf{B}$. Choosing **B** along the z-axis, $F_z = -\partial V/\partial z = \mu_z \partial B_z/\partial z$. Thus, atoms of opposite magnetic moment would have opposite trajectories.

The atoms exiting the furnace are randomly oriented; i.e., they have no preferred orientation of μ. Classically, one would expect that the atoms would have all possible magnetic moments between μ and $-\mu$, thus

---

[3] The experiment was later performed with hydrogen atoms, eliminating the concern that the silver closed shell might produce some residual orbital angular momentum.

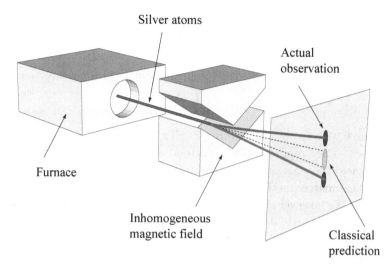

Fig. 15.2 Stern-Gerlach experiment. A beam of silver atoms is split in two beams after passing through an inhomogeneous magnetic field. This is in stark contradiction with the expected classical behavior.

producing a continuous smear on the detector. In fact, they produced only two smears, separated from each other. The only way to account for this was to assume an intrinsic angular momentum, "spin" which could assume only two orientations, up and down, $\pm z$.

Spin made its theoretical appearance in the construction of the ladder for orbital angular momentum, where, as discussed in Chapter 8, the possibility of half-integer quantum numbers appeared.

Keeping in mind that spin does not appear in the Schrödinger hamiltonian, and can therefore make no contribution to the non-relativistic energy, the experimental evidence for its presence must be accounted for [4]. This is accomplished by constructing a two-dimensional spin space, in which operators acting on the vectors representing spin orientations produce eigenvalues of $\pm 1/2$. We choose these to be along the z-axis, as per the Stern-Gerlach experiment.

In matrix formulation,

$$\chi_z^+ = \begin{pmatrix} 1 \\ 0 \end{pmatrix}, \quad \chi_z^- = \begin{pmatrix} 0 \\ 1 \end{pmatrix}.$$

Since we have chosen this basis set, the spin operator which yields $\pm 1/2$

---

[4]It is in fact a relativistic effect, appearing naturally in the solution of the Dirac equation.

must be

$$S_z = \frac{1}{2} \begin{pmatrix} 1 & 0 \\ 0 & -1 \end{pmatrix}.$$

The matrix itself is called σ_z, the Pauli spin matrix in the z direction:

$$\sigma_z \equiv \begin{pmatrix} 1 & 0 \\ 0 & -1 \end{pmatrix}.$$

We now wish to obtain the other Pauli spin matrices, σ_x and σ_y in this z-basis set. They represent the operators which acting on the vectors $\chi_x^{\pm}$ and $\chi_y^{\pm}$ yield the eigenvalues $\pm 1/2$ in those directions. This could be done by standard matrix methods, or alternately by using raising and lowering operators. We however shall obtain the vectors $\chi_x^{\pm}$ and $\chi_y^{\pm}$ from symmetry arguments, then write down σ_x and σ_y without deriving them, leaving it as a problem to demonstrate that they yield the correct values of spin.

Defining $\chi_x^+ \equiv a_{xz}^+ \chi_z^+ + a_{xz}^- \chi_z^-$, normalization yields

$$|a_{xz}^+|^2 + |a_{xz}^-|^2 = 1.$$

Similarly, $\chi_y^+ \equiv a_{yz}^+ \chi_z^+ + a_{yz}^- \chi_z^-$ yields

$$|a_{yz}^+|^2 + |a_{yz}^-|^2 = 1.$$

By symmetry and isotropy arguments χ_x^+ and χ_y^+ cannot be weighted favorably toward χ_z^+ or χ_z^-. Thus, $2|a_{xz}^+|^2 = 2|a_{yz}^+|^2 = 1$; i.e., $a_{xz}^{+*} a_{xz}^+ = a_{yz}^{+*} a_{yz}^+ = 1/2$. The possibilities (to within a sign) are

$$a_{xz}^+ = \pm a_{xz}^- = 1/\sqrt{2}, \ a_{yz}^+ = \pm a_{yz}^- = +i/\sqrt{2}.$$

This gives us all of the desired orthonormal eigenvectors:

$$\chi_x^+ = \frac{1}{\sqrt{2}} \begin{pmatrix} 1 \\ 1 \end{pmatrix}, \ \chi_x^- = \frac{1}{\sqrt{2}} \begin{pmatrix} 1 \\ -1 \end{pmatrix}$$

$$\chi_y^+ = \frac{i}{\sqrt{2}} \begin{pmatrix} 1 \\ i \end{pmatrix}, \ \chi_y^- = \frac{i}{\sqrt{2}} \begin{pmatrix} i \\ 1 \end{pmatrix}.$$

We now claim that

$$\sigma_x = \begin{pmatrix} 0 & 1 \\ 1 & 0 \end{pmatrix}$$

and

$$\sigma_y = \begin{pmatrix} 0 & -i \\ i & 0 \end{pmatrix}.$$

Problem 15.1. *Show that these give the correct values for spin in their respective directions.*

This must be the case because from the very beginning, the Stern-Gerlach experiment, the choice of which axis to call "z" was arbitrary. However, once having chosen z, we must choose x and y to preserve cyclic order: $\sigma_x \sigma_y = i \sigma_z$.

We said at the outset that the presence of additional lines in atomic spectra led to the discovery of spin. Spin cannot be derived from the Schrödinger equation [5]. Therefore, since it does not change the spatial wave functions, any contribution it makes to the total energy must be a small perturbation. The magnetic quantum numbers m_ℓ and m_s, introduced in the vector model of spin-orbit coupling are not good quantum numbers except in this approximation.

15.3   The Dirac Equation

Let us now turn to the development of the Dirac equation. We will begin with the consideration of a free particle, and then move on to a sketch of the solutions for the hydrogen atom. Although we shall not derive it, we emphasize at the outset that the Dirac equation predicts spin. Thus, spin-orbit coupling is a relativistic effect.

In special relativity, position and time play equivalent roles. Therefore, a relativistic equation must be symmetric in these. But the time dependent Schrödinger equation is linear in time and quadratic in position. Thus, it is necessary to make it either linear or quadratic in both.

The first attempts were to make the time dependence quadratic. Some led to interesting results. One case is the Klein-Gordon equation. A serious shortcoming was that it predicted a particle probability density that was not positive definite. However, it turned out to be useful for particles of spin zero.

Dirac went the other way, seeking a first-order equation in position and time; i.e., one which would explicitly contain $\partial/\partial t$ and $p = \nabla/i$. It should give the correct solution for a particle of any charge or none, as well as for a charged particle in electric and/or magnetic fields; viz., for both scalar and vector potentials. He looked for a linear form of the relativistic hamiltonian to match the linearity of the time-dependence. He accomplished this by

---

[5] In addition, its gyromagnetic ratio is twice the classical value.

writing the hamiltonian as

$$H = \boldsymbol{\alpha} \cdot \mathbf{p} + \beta m$$
$$= \alpha_1 p_1 + \alpha_2 p_2 + \alpha_3 p_3 + \beta m \ .$$

This is manifestly symmetric in the four-vector (p_1, p_2, p_3, E_0) where $E_0 = m$, thus yielding an equation which is first order in both ∇ and $\partial/\partial t$.

What then are the properties of α and β? They should be space- and time-independent, since $\mathbf{p}$ and m contain all of that information. They are also dimensionless.

We can now recapture $H\psi = E\psi$ by setting the relativistic form of the square of the energy, $p^2 + m^2$, equal to the square of the Dirac hamiltonian, taking care not to disturb the order of operations, since we do not know if α and β commute. Writing

$$p^2 + m^2 = (\boldsymbol{\alpha} \cdot \mathbf{p} + \beta m) \cdot (\boldsymbol{\alpha} \cdot \mathbf{p} + \beta m) \tag{15.4}$$

we obtain successively

$$p^2 + m^2 = \left(\sum_{i=1}^{3} \alpha_i p_i \right) \left(\sum_{j=1}^{3} \alpha_j p_j \right) + m^2 \beta^2$$
$$+ m \left[\left(\sum_{i=1}^{3} \alpha_i p_i \right) \beta + \beta \left(\sum_{i=1}^{3} \alpha_i p_i \right) \right]$$
$$= \left[\sum_{i=1}^{3} \alpha_i^2 p_i^2 + \sum_{j \neq i} \alpha_i \alpha_j p_i p_j \right] + m^2 \beta^2$$
$$+ m \left[\sum_{i=1}^{3} \alpha_i p_i \beta + \beta \sum_{i=1}^{3} \alpha_i p_i \right] \ . \tag{15.5}$$

Matching left and right sides,

$$p^2 = \sum_{i=1}^{3} \alpha_i^2 p_i^2 \Rightarrow \alpha_i^2 = 1 \ \text{ and } \ m^2 = \beta^2 m^2 \Rightarrow \beta^2 = 1.$$

The cross terms must separately vanish for each i, since β and each of the α_i's were all assumed to be independent of each other

$$0 = \alpha_i \alpha_j + \alpha_j \alpha_i, \ j \neq i$$

and

$$0 = \alpha_i \beta + \beta \alpha_i, \ i = 1, 2, 3.$$

Problem 15.2. *Prove these anti-commutation relations.*

These relations can be written in matrix form. A conventional choice is

$$\alpha_1 = \begin{pmatrix} 0\,0\,0\,1 \\ 0\,0\,1\,0 \\ 0\,1\,0\,0 \\ 1\,0\,0\,0 \end{pmatrix} \tag{15.6}$$

$$\alpha_2 = \begin{pmatrix} 0 & 0 & 0 & -i \\ 0 & 0 & i & 0 \\ 0 & -i & 0 & 0 \\ i & 0 & 0 & 0 \end{pmatrix} \tag{15.7}$$

$$\alpha_3 = \begin{pmatrix} 0 & 0 & 1 & 0 \\ 0 & 0 & 0 & -1 \\ 1 & 0 & 0 & 0 \\ 0 & -1 & 0 & 0 \end{pmatrix} \tag{15.8}$$

$$\beta = \begin{pmatrix} 1 & 0 & 0 & 0 \\ 0 & 1 & 0 & 0 \\ 0 & 0 & -1 & 0 \\ 0 & 0 & 0 & -1 \end{pmatrix} . \tag{15.9}$$

Writing the matrices as

$$\alpha_i = \begin{pmatrix} 0 & \sigma_i \\ \sigma_i & 0 \end{pmatrix} \tag{15.10}$$

$i = 1, 2, 3$

$$\beta = \begin{pmatrix} \mathbf{1} & 0 \\ 0 & -\mathbf{1} \end{pmatrix} \tag{15.11}$$

where the σ_i's are the Pauli spin matrices, $\mathbf{1} = \begin{pmatrix} 1\,0 \\ 0\,1 \end{pmatrix}$ and defining $\boldsymbol{\sigma_i} \equiv \begin{pmatrix} \sigma_i & 0 \\ 0 & \sigma_i \end{pmatrix}$ simplifies calculation of the anti-commutation rules.

Putting all of this together, the Dirac equation for a free particle is

$$H\psi = i\frac{\partial \psi}{\partial t} = E\psi = (\boldsymbol{\alpha} \cdot \mathbf{p} + \beta m)\psi = \left(\boldsymbol{\alpha} \cdot \frac{\nabla}{i} + \beta m \right) \psi. \tag{15.12}$$

But since the α_i and β_i are 4×4 matrices, ψ must be a 4-component column vector:

$$\psi = \begin{pmatrix} \psi_1 \\ \psi_2 \\ \psi_3 \\ \psi_4 \end{pmatrix} .$$

For a particle of positive charge q, we can introduce an electromagnetic field with vector potential $\mathbf{A}$ and scalar potential Φ, and any additional potential V. With $\mathbf{p} \to \mathbf{p} - q\mathbf{A}$ [6], the Dirac hamiltonian becomes

$$H = \boldsymbol{\alpha} \cdot (\mathbf{p} - q\mathbf{A}) + \beta m + q\Phi + V.$$

If none of these potentials is time-dependent, we can separate time out, just as for the Schrödinger equation, and obtain the time-independent Dirac equation. $\Psi(\mathbf{r}, t) = \psi(\mathbf{r})e^{-iEt}$, where $\psi(\mathbf{r})$ obeys:

$$H\psi = [\boldsymbol{\alpha} \cdot (\mathbf{p} - q\mathbf{A}) + \beta m + q\Phi + V]\psi = E\psi. \qquad (15.13)$$

For no external electromagnetic fields, but only a scalar potential V, the Dirac equation becomes

$$H\psi = [\boldsymbol{\alpha} \cdot \mathbf{p} + \beta m + V]\psi = E\psi \qquad (15.14)$$

where $V = -e^2/r$ for the hydrogen atom.

15.4   Constants of the Motion

As in the non-relativistic case, we look for an angular momentum operator which commutes with H. This is the criterion for conservation of that momentum. In Chapter 3 we showed that an operator which commutes with H is conserved. Its expectation values are time-independent and consequently its eigenvalues are good quantum numbers. We note that, just as in the case of the Schrödinger equation, the results we will obtain are the same for the free particle and the hydrogen atom, since the hydrogen potential has no angular dependence. This is the case because $\mathbf{L}$ commutes[7] with r^2 . From this we can conclude that $\mathbf{L}$ commutes with any scalar central potential $V(|\mathbf{r}|)$.

We first try the orbital angular momentum $\mathbf{L} = \mathbf{r} \times \mathbf{p}$. Since it operates in ordinary 3-space, it commutes with the α_i's but not the p_i's. The commutator of $\mathbf{L}$ with $\mathbf{p}$ was given in Problem 8.1.

Using these results, we find that $[\mathbf{L}, H]$ does not vanish; in fact it equals $i(\boldsymbol{\alpha} \times \mathbf{p})$.

Problem 15.3. *Prove this. (Show for one component, and generalize.)*

---

[6]This is known as minimal substitution.

[7]The proof is tedious but straightforward, and may be found in many introductory quantum mechanics textbooks.

We must then look for an additional angular-momentum-like operator whose commutator with H is $-i(\boldsymbol{\alpha} \times \mathbf{p})$ which we can add to $\mathbf{L}$ to obtain an operator which does commute with H. We have already seen that the Dirac hamiltonian's α matrices are the result of relativistic extension of the Schrödinger equation, and they are related to the Pauli spin matrices. We therefore examine these $\boldsymbol{\alpha}$'s, and find that

$$[\boldsymbol{\sigma}, H] = -2i(\boldsymbol{\alpha} \times \mathbf{p}).$$

Problem 15.4. *Show this.*

Since $\mathbf{L} + \boldsymbol{\sigma}/2$ commutes with H, we can define a spin angular momentum $\mathbf{S} \equiv \boldsymbol{\sigma}/2$, and a total angular momentum $\mathbf{J} \equiv \mathbf{L} + \mathbf{S}$, which is conserved. We thus see how spin arises naturally from the Dirac equation.

Of what operators in addition to the hamiltonian is ψ an eigenfunction? That is, which of their quantum numbers are good? The magnitude of the angular momentum vector is not: $[L^2, H] \neq 0$. As we noted earlier, all of the components of $\mathbf{L}$ commute with α, β, m, and V. So we need only consider the commutation of L^2 with the components of $\mathbf{p}$. We examine the commutator for one component of L^2 with each component p_i, say $[L_x^2, p_i]$

From the results of Problem 8.1, we find that $[L_x^2, p_y] = i(L_x p_z + p_z L_x)$. Since L_x appears explicitly in this result, it cannot be cancelled by any other component: the components of L do not commute. So L^2 is not conserved.

S^2 does commute with the hamiltonian. We expect this, since spin is an intrinsic property of the particle, and therefore its magnitude should not change. To confirm this, we observe that the square of each of the Pauli matrices is the identity operator, which commutes with everything.

Since $\mathbf{L}$, $\mathbf{S}$, and L^2 do not commute with H, they are not conserved. Therefore, m_ℓ, m_s and ℓ are not good quantum numbers. Thus, the analysis of spin-orbit coupling and its contribution to the fine structure is only correct in the low velocity limit, where, as we discussed, it is a perturbation, not included in the Schrödinger hamiltonian. This leaves only s, j, and m_j as good quantum numbers. We therefore expect them to appear as indices in the Dirac wave function.

Returning to consideration of that wave function, we redefine

$$\psi \equiv \begin{pmatrix} \Phi \\ \Omega \end{pmatrix} \quad \text{where } \Phi \equiv \begin{pmatrix} \psi_1 \\ \psi_2 \\ 0 \\ 0 \end{pmatrix} \text{ and } \Omega \equiv \begin{pmatrix} 0 \\ 0 \\ \psi_3 \\ \psi_4 \end{pmatrix}. \tag{15.15}$$

In order to understand the overall structure of the solutions to the Dirac equation, we will start with the simple case of the free particle. We will then move on to the hydrogen atom.

Since α_i and β commute with p_i, we can first operate on the components ψ_i with them, and subsequently with the p_i's to obtain the solutions to the Dirac equation. Using the matrix forms of α_i and β, we obtain

$$\alpha_1 \begin{pmatrix} \psi_1 \\ \psi_2 \\ \psi_3 \\ \psi_4 \end{pmatrix} = \begin{pmatrix} \psi_4 \\ \psi_3 \\ \psi_2 \\ \psi_1 \end{pmatrix}$$

$$\alpha_2 \begin{pmatrix} \psi_1 \\ \psi_2 \\ \psi_3 \\ \psi_4 \end{pmatrix} = \begin{pmatrix} -i\psi_4 \\ i\psi_3 \\ -i\psi_2 \\ i\psi_1 \end{pmatrix}$$

$$\alpha_3 \begin{pmatrix} \psi_1 \\ \psi_2 \\ \psi_3 \\ \psi_4 \end{pmatrix} = \begin{pmatrix} \psi_3 \\ -\psi_4 \\ \psi_1 \\ -\psi_2 \end{pmatrix}$$

$$\beta \begin{pmatrix} \psi_1 \\ \psi_2 \\ \psi_3 \\ \psi_4 \end{pmatrix} = \begin{pmatrix} \psi_1 \\ \psi_2 \\ -\psi_3 \\ -\psi_4 \end{pmatrix} .$$

Placing these results into the Dirac equation, with $p_i = -i\,\partial/\partial x_i$, we obtain a set of four coupled differential equations:

$$(-i\,\partial\psi_4/\partial x - \partial\psi_4/\partial y) - i\,\partial\psi_3/\partial z + m\psi_1 = E\psi_1$$

$$(-i\,\partial\psi_3/\partial x + \partial\psi_3/\partial y) + i\,\partial\psi_4/\partial z + m\psi_2 = E\psi_2$$

$$\tag{15.16}$$

$$(-i\,\partial\psi_2/\partial x - \partial\psi_2/\partial y) - i\,\partial\psi_1/\partial z - m\psi_3 = E\psi_3$$

$$(-i\,\partial\psi_1/\partial x + \partial\psi_1/\partial y) + i\,\partial\psi_2/\partial z - m\psi_4 = E\psi_4 .$$

We can gain insight into the meaning of these equations by choosing the momentum in the z-direction. Then the first two terms in each equation disappear. We can further, for a free particle, take the momentum as a fixed quantity, the eigenvalue of $-i\,\partial\psi_i/\partial z = p\,\psi_i$, and substitute p for $-i\,\partial/\partial z$. Then

$$p\psi_3 - (E - m)\psi_1 = 0$$
$$-p\psi_4 - (E - m)\psi_2 = 0$$
$$p\psi_1 - (E + m)\psi_3 = 0 \qquad (15.17)$$
$$-p\psi_2 - (E + m)\psi_4 = 0 .$$

The equations are decoupled into two pairs: ψ_1, ψ_3 and ψ_2, ψ_4. The two equations involving the former produce the determinantal equation

$$\begin{vmatrix} E - m & -p \\ -p & E - m \end{vmatrix} = 0$$

whence $E = \pm\sqrt{p^2 + m^2}$. (The same result obtains using ψ_2, ψ_4.)
We also find

$$\psi_4 = (E - m)\psi_2/p$$

and

$$\psi_3 = p\psi_1/(E + m).$$

Substituting $p = \pm\sqrt{E^2 - m^2}$,

$$\psi_3 = \sqrt{(E - m)/(E + m)}\ \psi_1$$

and

$$\psi_4 = \sqrt{(E - m)/(E + m)}\ \psi_2.$$

In the highly relativistic case $E \gg m$, ψ_3 and ψ_4 are comparable in size to ψ_1 and ψ_2. In less extreme situations, ψ_3 and ψ_4 are smaller than ψ_1 and ψ_2 (For $v = c/2$, $\Omega = 0.27\Phi$.) The former have come to be known as the small parts of the wave function and the latter, the large parts. Eq. (15.15) divided ψ neatly into these components.

At this point we have learned all we need from studying the free particle, and we move on to explicit consideration of the hydrogen atom. As $v \to 0$, the Dirac wave function must reduce to the Schrödinger wave function with the spin tacked on. We can separate the radial part of the wave function, since only it depends on the potential, from the rest, which is now analogous to the spherical harmonics, but with spin explicitly included.

$$\psi \equiv \begin{pmatrix} \Phi \\ \Omega \end{pmatrix} \sim \frac{1}{r} \begin{pmatrix} G_n(r)\mathcal{Y}^m_{\ell\,j}(\theta,\phi,s) \\ i\,F_n(r)\mathcal{Y}^m_{\ell'\,j}(\theta,\phi,s) \end{pmatrix}. \tag{15.18}$$

G is the large part of the radial wave function; F is the small part [8]. As speed decreases, G should reduce to R, the radial part of the Schrödinger equation, F should vanish, and the $\mathcal{Y}$'s should reduce to the non-relativistic products of the spherical harmonics and the spin functions. Finally, parity of the wave functions [9] must be preserved as we pass from the relativistic to the non-relativistic case. $\ell' = \ell \pm 1$ depending on j: $\ell \equiv j + \tilde{\omega}/2$, $\ell' \equiv j - \tilde{\omega}/2$. $\tilde{\omega}$: $\tilde{\omega} = \pm 1$ for states of parity $(-)^{j \pm 1/2}$. m is m_ℓ: $-\ell \le m_\ell \le \ell$.

We must tread cautiously when examining the angular-spin functions. We can only describe the wave function with quantum numbers s, j, and m_j. For the single electron of hydrogen, $s = 1/2$ and j is a half-integer. But the spherical harmonics, and anything that reduces to them, have indices m_ℓ and ℓ, not good quantum numbers. Yet they appear in the wave functions. How can this be? Actually, it's simply an accounting scheme, labeling states $|s\ j\ m_j\rangle$ by all possible combinations of values of integers ℓ and m_ℓ for $j \pm 1/2$, $-j \le m_j \le j$. Let us sketch how this works. For the Schrödinger equation for hydrogen, the only good quantum numbers (ignoring the principal quantum number) are ℓ and m_ℓ. Only the ad hoc addition of spin introduces the spin-orbit-coupling perturbation and the total angular momentum $\mathbf{J} = \mathbf{L} + \mathbf{S}$, with concomitant quantum numbers s, m_s, j, and m_j, and the rule $j = \ell \pm 1/2$. For the Dirac equation, even though ℓ is not a good quantum number, it can still be formally defined: $\ell \equiv j \mp 1/2$. We denote the possibilities as ℓ and ℓ'.

If we take all possibilities into account, we obtain two sets of 4-component wave functions.

$$\psi \equiv \begin{pmatrix} \Phi \\ \Omega \end{pmatrix} = \begin{pmatrix} \psi_1 \\ \psi_2 \\ \psi_3 \\ \psi_4 \end{pmatrix},$$

one for $j = \ell + 1/2$ and one for $j = \ell - 1/2$, with the proviso that $\ell' \ne \ell$. All of this is tedious but straightforward, and true for any central potential. Thus, for example, if $j = 3/2$, then ℓ and ℓ' can be either 1 or 2. We

---

[8]The i multiplying F appears because of the connection of large and small components through $p = -i\partial/\partial x$ in Eq. (15.16).

[9]$\psi(-\mathbf{r}) = \pm\psi(\mathbf{r})$.

saw that the fine structure, a combination of the lowest order relativistic effect and the spin-orbit coupling, each of which had an ℓ dependence, itself depended only on j. This is the cause. Thus, for $j = 3/2$, the states $^2P_{3/2}$ and $^2D_{3/2}$ have the same energy [10]. Similarly, $m_\ell = m_j \pm 1/2$, although neither m_ℓ nor the spin projection $m_s = \pm 1/2$ is a good quantum number.

To summarize, the good quantum numbers are j, m_j, and $\tilde{\omega}$, which has determined whether $\ell = j \pm 1/2$. Schematically,

$$\mathcal{Y}^{m_\ell = m_j \pm \frac{1}{2}}_{\tilde{\omega}, \, \ell = j \pm 1, \, j} \left(\theta, \phi, \frac{1}{2} \right) \longrightarrow Y_\ell^m (\theta, \phi) \, \chi \left(\frac{1}{2}, \pm \frac{1}{2} \right)$$

with positive or negative parity.

Although we have discussed spin at some length, we have not proved that the Dirac equation predicts it [11].

Let us now obtain the radial functions for hydrogen. Since the Dirac equation is a first order differential equation, the result of a good deal of algebra [12] is a pair of coupled first order equations in G and F:

$$\left(-\frac{d}{d\rho} + \frac{\tau}{\rho} \right) F = \left(-\nu + \frac{\gamma}{\rho} \right) G \tag{15.19}$$

$$\left(\frac{d}{d\rho} + \frac{\tau}{\rho} \right) G = \left(\frac{1}{\nu} + \frac{\gamma}{\rho} \right) F . \tag{15.20}$$

The constants are defined as follows.

$$\kappa \equiv \sqrt{m^2 - E^2}$$

$$\rho \equiv \kappa r$$

$$\nu \equiv \left(\sqrt{\frac{m - E}{m + E}} \right)$$

$$\tau \equiv \left[j + \frac{1}{2} \right] \tilde{\omega}$$

$$\gamma \equiv Ze^2 .$$

γ is just $Z\alpha$, allowing for calculation of the fine structure for hydrogenic atoms.

---

[10]Technically, these are levels, not states. The latter include the magnetic quantum number m_j, and are degenerate in the absence of an external magnetic field.
[11]For a proof, see A. Messiah, *Quantum Mechanics*, v.2, Paris: Dunod Publ. (1964).
[12]H. Bethe and E.E. Salpeter, *Quantum Mechanics of One- and Two- Electron Atoms*. NY: Springer (1977).

Note also that for real κ and ν, $E < m$. As we know, for bound states such as those we are investigating, the contribution to the energy from the potential is negative. Then the total energy; viz., the bound state plus rest energy m must be less than the positive rest energy m. From Dirac's conclusion that there is a sea of negative rest energy states below $-m$, the symmetry of his positive and negative solutions constrains the total energy even more: $-m \leq E \leq +m$.

We now carry out the same "peel-off + series expansion + truncation" procedure as we did for the Schrödinger equation for hydrogen.

$$G \sim \rho^s \, e^{-\rho} \, \Sigma_{i=0}^{\infty} \, a_i \rho^i$$

$$F \sim \rho^s \, e^{-\rho} \, \Sigma_{i=1}^{\infty} \, b_i \rho^i$$

We substitute these into the pair of coupled differential equations, and match terms of the same order, giving us recursion relations between successive terms. In order for the wave functions to be normalizable, we must truncate the series for G and F. This will gives us the allowed energies.

$$E_{nj} = m \left[1 + \left(\frac{\alpha}{n - (j + \frac{1}{2}) + \sqrt{(j + \frac{1}{2})^2 - \alpha^2}} \right)^2 \right]^{-\frac{1}{2}} \tag{15.21}$$

This expression includes the rest energy m.

Problem 15.5. *Prove that the exact result exclusive of the rest energy, expanded to order α^4 produces the Bohr result plus the fine structure*

$$E_{nj} = E_n \left\{ \frac{1}{n^2} + \frac{\alpha^2}{n^4} \left[\frac{n}{j + 1/2} - \frac{3}{4} \right] \right\}.$$

E_n is the Bohr energy.

Note that this approximation does not reintroduce the "bad" quantum number ℓ.

We now see that the ℓ-degeneracy was not an artifact of the approximate fine structure calculation, but is true for the exact case. The equal energy of states of the same j in Figure 15.1 comes from the Dirac equation.

15.5   SUSYQM

What does all of this have to do with SUSYQM? We have seen two suggestive bits of evidence. The first, apparent in Figure 15.1 and in the Dirac energy eigenvalues, is the appearance of adjacent columns with equal energies for a given n and j, except for the lowest energy. These are due to the same (Coulomb) potential, yet they look a lot like the energies of partner potentials.

The second piece of evidence is in the pair of coupled equations for F and G in Eqs. (15.19) and (15.20). The operation $\left(-\frac{d}{d\rho} + \frac{\tau}{\rho}\right) F$ yields $\left(-\nu + \frac{\gamma}{\rho}\right) G$ while the operation $\left(\frac{d}{d\rho} + \frac{\tau}{\rho}\right) G$ yields $\left(\frac{1}{\nu} + \frac{\gamma}{\rho}\right) F$. This looks suspiciously like raising and lowering operations associated with a super-potential of form τ/ρ.

But here we encounter a problem. The eigenfunctions of partner hamiltonians should be of the form $A^+ F \sim G$, $A^- G \sim F$ where all of the ρ (that is, r) dependence resides on the left side, in $A^\pm$. This is not the case in Eqs. (15.19) and (15.20). There is a dependence on the right hand sides of form F/ρ and G/ρ.

We can rewrite Eqs. (15.19) and (15.20) in a more compact matrix form:

$$\begin{pmatrix} dG/dr \\ dF/dr \end{pmatrix} + \frac{1}{r} \begin{pmatrix} k & -\gamma \\ \gamma & -k \end{pmatrix} \begin{pmatrix} G \\ F \end{pmatrix} = \begin{pmatrix} (m+E)F \\ (m-E)G \end{pmatrix}.$$

Here we see the problem: the second term on the left is mixing F and G. If it can be diagonalized, then the equations can be recast in SUSYQM form. A solution [13] is to apply a linear transformation of the functions G, F into $\mathcal{G}, \mathcal{F}$:

$$\begin{pmatrix} k+s & -\gamma \\ \gamma & k+s \end{pmatrix}$$

$$\begin{pmatrix} \mathcal{G} \\ \mathcal{F} \end{pmatrix} = D \begin{pmatrix} G \\ F \end{pmatrix}.$$

Substituting $\gamma \equiv \sqrt{(k^2 - s^2)}$ yields the appropriate supersymmetric form of the coupled equations:

$$\left(\frac{k}{s} + \frac{m}{E}\right) \mathcal{F} = \left(\frac{d}{d\mu} + \frac{s}{\mu} - \frac{\gamma}{s}\right) \mathcal{G} \qquad (15.22)$$

$$\left(\frac{k}{s} - \frac{m}{E}\right) \mathcal{G} = \left(\frac{-d}{d\mu} + \frac{s}{\mu} - \frac{\gamma}{s}\right) \mathcal{F} \qquad (15.23)$$

where $\mu \equiv Er$.

[13] A. Sukumar, "Supersymmetry and the Dirac equation for a central Coulomb field", *J. Phys.* **A18** L697 (1985).

This is precisely the form needed for supersymmetric partner eigenfunctions, related by $A^{\pm} = \left(\pm \frac{d}{d\mu} + \frac{s}{\mu} - \frac{\gamma}{s} \right)$. We can then decouple the equations, yielding the eigenvalue equations for $\mathcal{G}$ and $\mathcal{F}$:

$$A^{-}A^{+}\mathcal{F} \equiv H_{+}\mathcal{F} = \left(\frac{\gamma^2}{s^2} + 1 - \frac{m^2}{E^2} \right)\mathcal{F} = \left(\frac{k^2}{s^2} - \frac{m^2}{E^2} \right)\mathcal{F} \qquad (15.24)$$

$$A^{+}A^{-}\mathcal{G} \equiv H_{-}\mathcal{G} = \left(\frac{\gamma^2}{s^2} + 1 - \frac{m^2}{E^2} \right)\mathcal{G} = \left(\frac{k^2}{s^2} - \frac{m^2}{E^2} \right)\mathcal{G}. \qquad (15.25)$$

Thus, the hamiltonians H_{-} and H_{+} share common eigenvalues, except for the extra case where $A^{-}\mathcal{G} = 0$, which yields the ground state wave function $\mathcal{G}(0) \sim \mu^s e^{-\mu/s}$ with eigenvalue $E^{(0)} = \frac{m^2}{k^2/s^2}$.

Note that neither H_{-} and H_{+} is the Dirac hamiltonian, so their eigenvalues E are not the hydrogen atom energy levels. Those will be obtained from the terms $\frac{m^2}{E^2}$

The application of the differential forms of $A^{\pm}$ in Eqs. (15.24) and (15.25) yields

$$\left[-\frac{d^2}{d\mu^2} - 2\frac{\gamma}{\mu} - \frac{s(s-1)}{\mu^2} + 1 - \frac{m^2}{E^2} \right]\mathcal{F} = 0 \qquad (15.26)$$

$$\left[\frac{d^2}{d\mu^2} + 2\frac{\gamma}{\mu} - \frac{s(s+1)}{\mu^2} + 1 - \frac{m^2}{E^2} \right]\mathcal{G} = 0 \qquad (15.27)$$

where $\mu \equiv Er$.

Problem 15.6. *Obtain equations (15.26) and (15.27) from equations (15.24) and (15.25).*

From $A^{\pm} = \left(\pm \frac{d}{d\mu} + \frac{s}{\mu} - \frac{\gamma}{s} \right)$ we can obtain the superpotential from either $A^{-}A^{+}$ or $A^{+}A^{-}$. The former gives us $A^{-}A^{+} = H_{+} = W^2 + dW/d\mu$. We compare this with

$$A^{-}A^{+} = -\frac{d^2}{d\mu^2} + \frac{s^2}{\mu^2} - \frac{s}{\mu^2} + \frac{\gamma^2}{s^2} - 2\frac{\gamma}{s}.$$

By inspection, $s^2/\mu^2 + \gamma^2/s^2 - 2\gamma/s = (s/\mu - \gamma/s)^2$. We try $W = s/\mu - \gamma/s$. Then $dW/d\mu = -s/\mu^2$. So indeed $W = s/\mu - \gamma/s$.

Now we see that we actually have the equivalent of a system of partner potentials $W^2 \pm \frac{dW}{d\mu}$.

But we can do even better. From Eqs. (15.26) and (15.27) we see that the simple replacement of s by $s+1$ in the first yields the second. In other words, we have shape invariance.

$$H_+(\mu, s, \gamma) = H_-(\mu, s+1, \gamma) + \frac{\gamma^2}{s^2} - \frac{\gamma^2}{(s+1)^2} \qquad (15.28)$$

We now know that each of the columns in Figure 15.1 can be obtained without recourse to any other. The eigenvalues are

$$E_n^{(-)} = E_{n-1}^{(+)} = \gamma^2 \left[\frac{1}{s^2} - \frac{1}{(s+1)^2} \right] \qquad (15.29)$$

Finally, we relate the actual Dirac fine structure energy spectrum of hydrogen E_n to the spectrum $E_n^{(-)}$ from Eqs. (15.24), (15.25) (where it is called simply E) and Eq.(15.29).

$$\left(\frac{k^2}{s^2} - \frac{m^2}{E_n^2} \right) \equiv E_n^{(-)}.$$

This yields

$$E_n = \frac{m}{\sqrt{1 + \frac{\gamma^2}{(s+n)^2}}}, \quad n = 0, 1, 2, \ldots . \qquad (15.30)$$

Putting $\hbar$ and c back in, and setting $Z = 1$ for hydrogen yields $\gamma = \alpha = e^2/\hbar c$. We obtain the expression for the fine structure energy from Eq. (15.21).

$$E_{nj} = m \left\{ \left[1 + \left(\frac{\alpha}{n - (j + \frac{1}{2}) + \sqrt{(j + \frac{1}{2})^2 - \alpha^2}} \right)^2 \right]^{-\frac{1}{2}} \right\}. \qquad (15.31)$$

The only mismatch is due to the fact that n starts at 1 in Eq. (15.21) while in Eq. (15.31), n starts at 0, as per the SUSY convention.

We have thus used SUSYQM to obtain the exact fine structure spectrum of hydrogen, to account for the degeneracy in j, to show that the appearance of adjacent partner-potential-like ladders in Figure 15.1 is no accident, and finally to make good on our claim that the Dirac equation is "intrinsically" supersymmetric.

Chapter 16

Scattering in SUSYQM

Scattering plays a very important role in the probing and modeling of atomic and sub-atomic interactions. Our early understanding of the structure of atoms came from scattering experiments. Also in nuclear physics, much of the information available was obtained from scattering of judiciously chosen projectiles by nuclei.

In scattering we typically study a very large ensemble of "free" projectiles hurled towards the center of a localized interaction. In this chapter we will study scattering by interactions such as one dimensional potential barriers (or wells), and three dimensional central potentials. In one dimension, as a result of the interaction with a potential barrier, some particles are reflected by it, and others are transmitted. For three dimensional cases, we will only consider potentials that are spherically symmetric. For such interactions, it suffices to look at the radial part of the system, as the angular parts of the wave function are given by the spherical harmonics.

In one dimensional scattering, a variable of interest is the ratio of particles reflected and transmitted to the number of incident particles, observed very far from the localized potential. For analysis of three dimensional scattering, we will consider the z-axis along the incident beam and place the origin at the center of the potential; i.e., the scatterer. Our objective will be to predict the flux of particles in all directions as given by coordinates θ, ϕ. As we will see later, our quantity of interest will be the scattering amplitude.

As we have seen in previous chapters, complete information about the bound states of a shape invariant system; i.e., its eigenvalues and eigenfunctions, can be found using the SUSYQM formalism. For scattering states too, supersymmetric quantum mechanics provides new insights. In this chapter, we will learn about scattering states using SUSYQM and shape invariance alone.

16.1   The One Dimensional Case

In one dimension, for particles to be free far away from a localized potential, the potential must tend towards a finite asymptotic value as $x \to -\infty$, or $x \to \infty$, or both. Let us call these finite values $V_{\pm\infty}$. We assume that incident free particles approach the potential from the left, and hence are described by a plane wave[1] e^{ikx}. In such cases, the asymptotic limits of the wave functions with energy E are given by

$$\lim_{x \to -\infty} \psi(x, E) \sim e^{ikx} + r(k)\, e^{-ikx} \quad,$$

and

$$\lim_{x \to \infty} \psi(x, E) \sim t(k')\, e^{ik'x} \,, \tag{16.1}$$

where k and k' are the asymptotic momenta of the particle. They are related to the energy by $E = k^2 + V_{-\infty} = k'^2 + V_{+\infty}$. Reflection and transmission amplitudes $r(k)$ and $t(k)$ are related respectively to the coefficient of reflection by $\mathcal{R} = |r(k)|^2$ and the coefficient of transmission by $\mathcal{T} = |t(k)|^2$. These amplitudes are measures of relative strengths of the outgoing waves, ($r(k)\, e^{-ikx}$ and $t(k')\, e^{ik'x}$), compared to the incident waves e^{ikx} from the left.

In supersymmetric quantum mechanics, since we have two partner potentials $V_{\pm} \equiv W^2(x) \pm W'(x)$, we will have to consider two different asymptotic limits of the type given in Eq. (16.1), one for each potential. Since the potentials reach finite asymptotic values at $\pm\infty$, we assume that so does the superpotential $W(x)$; i.e., $\left.\frac{dW(x)}{dx}\right|_{\pm\infty} = 0$. Thus, for a SUSYQM system, we obtain

$$\lim_{x \to -\infty} \psi^{\mp}(x, E) \sim e^{ikx} + r^{\mp}(k)\, e^{-ikx} \quad;$$

and

$$\lim_{x \to \infty} \psi^{\mp}(x, E) \sim t^{\mp}(k')\, e^{ik'x} \,. \tag{16.2}$$

Momenta k and k' are given by $\sqrt{E - (W_{-\infty})^2}$ and $\sqrt{E - (W_{+\infty})^2}$, where $W_{\pm\infty}$ are the asymptotic values of the superpotential as $x \to \pm\infty$.

---

[1]Note that this term is multiplied by a time-dependent term $e^{-i\omega t}$ and thus incident particles are appropriately represented by $e^{i(kx-\omega t)}$, a plane wave moving toward increasing x.

Now, using SUSYQM algebra, we will show that r^+ and r^-, and similarly t^+ and t^- are related to each other. The wave functions of H_+ and H_- are related by

$$\left(\frac{d}{dx} + W(x)\right) \psi_E^-(x) = N_1 \, \psi_E^+(x)$$

$$\left(-\frac{d}{dx} + W(x)\right) \psi_E^+(x) = N_2 \, \psi_E^-(x) \, ,$$

(16.3)

where N_1 and N_2 are constants. Applying these conditions to the asymptotic forms given in Eq. (16.2), we obtain

$$\left(\frac{d}{dx} + W_{-\infty}\right) \left(e^{ikx} + r^-(k)\, e^{-ikx}\right) = N_1 \left(e^{ikx} + r^+(k)\, e^{-ikx}\right) \, ,$$

which gives

$$(ik + W_{-\infty})\, e^{ikx} + r^- \left(-ik + W_{-\infty}\right) e^{-ikx} = N_1 \left(e^{ikx} + r^+ e^{-ikx}\right) \, ,$$

(16.4)

where we have suppressed the k-dependence of $r^\pm$. Similarly, we obtain

$$(-ik + W_{-\infty})\, e^{ikx} + r^+ \left(ik + W_{-\infty}\right) e^{-ikx} = N_2 \left(e^{ikx} + r^-\, e^{-ikx}\right) \, .$$

(16.5)

Comparing the coefficients of $e^{\pm ikx}$ in Eq. (16.4), we obtain following relationships:

$$(ik + W_{-\infty}) = N_1 \quad \text{and} \quad r^- \left(-ik + W_{-\infty}\right) = N_1\, r^+. \quad (16.6)$$

Solving these two equations, we obtain

$$r^+ = \left(\frac{W_{-\infty} - ik}{W_{-\infty} + ik}\right) r^- \; ; \qquad r^- = \left(\frac{W_{-\infty} + ik}{W_{-\infty} - ik}\right) r^+ \, . \quad (16.7)$$

This identity, which follows from SUSYQM, has profound impact. Because $\left(\frac{W_{-\infty} + ik}{W_{-\infty} - ik}\right)$ is just a phase, the coefficients of reflection $\mathcal{R}^\pm \equiv |r^\pm(k)|^2$, are the same for partner potentials $V_\pm$. If $r^-(k) = 0$, then $r^+(k) = 0$ - we will see that this leads to the reflectionless potentials. It also shows that if $W_{-\infty}$ is negative, then $r^-(k)$ has an additional pole (compared to $r^+(k)$) on the positive imaginary axis of the k-plane. This pole corresponds to the additional zero-energy eigenstate of H_-. A similar analysis leads to

$$t^+(k) = \left(\frac{W_{-\infty} - ik}{W_{+\infty} - ik'}\right) t^-(k) \, . \quad (16.8)$$

Problem 16.1. *Prove the identity claimed in Eq. (16.8).*

Problem 16.2. *Show that Eq. (16.5) also generates the same relationship between $r^-(k)$ and $r^+(k)$ as given by Eq. (16.7).*

Problem 16.3. *Consider a system described by the a shape invariant superpotential $W(x, a_0) = a_0 \tanh x$.*

(1) Determine the partner potentials $V_\pm(x, a_0)$ generated by this superpotential.

(2) Using plotting software, plot superpotential $W(x, a_0)$ and partner potentials $V_\pm(x, a_0)$ for $a_0 = 1$ and $-5 < x < 5$.

(3) For $a_0 = 1$, determine the coefficients $r^-(k, a_0)$ and $t^-(k, a_0)$.

So far we have been focused on the implications of supersymmetry. Now let us turn to potentials with shape invariance. For such potentials, we have

$$V_+(x, a_i) = V_-(x, a_{i+1}) + R(a_0). \tag{16.9}$$

Since these two potentials, $V_+(x, a_i)$ and $V_-(x, a_{i+1})$, differ only by a constant term $R(a_0)$, they share the same scattering states as well as the bound states. This implies that they must also have the same reflection (and transmission) amplitudes. I.e., $r^+(k, a_i) = r^-(k, a_{i+1})$ and $t^+(k, a_i) = t^-(k, a_{i+1})$. Hence, we obtain the following recursion relation,

$$r^-(k, a_1) = \left(\frac{W_{-\infty}(a_0) - ik}{W_{-\infty}(a_0) + ik} \right) r^-(k, a_0). \tag{16.10}$$

At this point we will consider all translational shape invariant potentials that allow for scattering[2] and explore the consequences of the above recursion relation. But we must note at the outset that we cannot solve the problem entirely without eventual recourse to the Schrödinger equation.

(1) Scarf II Potential: The superpotential for this system is given by

$$W(x, A, B) = A \tanh x + B \operatorname{sech} x. \tag{16.11}$$

The asymptotic value of the superpotential W at $-\infty$ is $-A$. Hence, Eq. (16.10) leads to

$$r^-(k, A + 1, B) = \frac{(A + ik)}{(A - ik)} r^-(k, A, B). \tag{16.12}$$

Since any function can be written as a ratio of two functions, albeit there are an infinitely large number of ways to do so, let us write

---

[2]Shape invariant potentials such as harmonic oscillators have no upper bounds. As a result, such potentials have no scattering states.

$r^-(k, A, B)$ as the ratio of two functions $u^-(k, A + ik, B)$ and $v^-(k, A - ik, B)$. Thus,

$$r^-(k, A, B) = \frac{u^-(k, A + ik, B)}{v^-(k, A - ik, B)} \equiv \frac{u^-(k, A', B)}{v^-(k, A'', B)} , \qquad (16.13)$$

where $A' \equiv A + ik$ and $A'' \equiv A - ik$. Eq. (16.12) now implies

$$\frac{u^-(k, A' + 1, B)}{v^-(k, A'' + 1, B)} = \frac{A'}{A''} \frac{u^-(k, A', B)}{v^-(k, A'', B)}. \qquad (16.14)$$

This is a difference equation, akin to the recursion relations for special functions of mathematical physics. To solve it, let us compare it with a well known property of Gamma functions[3] ; viz., $\Gamma(z + 1) = z\,\Gamma(z)$. If a function $f(z)$ obeys $f(z + 1) = z\,f(z)$, we have $\frac{f(z+1)}{\Gamma(z+1)} = \frac{f(z)}{\Gamma(z)}$ for all values of the variable z. This is possible only if these ratios are equal to a function $\alpha(z)$ with period one; i.e., $\alpha(z + 1) = \alpha(z)$. Employing this property, we see that the dependence of $u^-(k, A', B)$ and $v^-(k, A'', B)$ on A' and A'' should be proportional to $\Gamma(A')$ and $\Gamma(A'')$ respectively. Thus, we can write

$$u^-(k, A', B) = \Gamma(A') \times \alpha_k^u(A, B)$$

and

$$v^-(k, A'', B) = \Gamma(A'') \times \alpha_k^v(A, B),$$

where $\alpha_k^u(A, B)$ and $\alpha_k^v(A, B)$ are functions of A, B, and k. The dependence on A must be a periodic function of period unity; i.e., $\alpha_k^{u,v}(A + 1, B) = \alpha_k^{u,v}(A, B)$. Putting all of the above together, we obtain

$$r^-(k, A, B) = \frac{\Gamma(A + ik)}{\Gamma(A - ik)} \alpha_k(A, B) \qquad (16.15)$$

where α_k, as stated earlier, is a function of k, A and B.

This is as far as we can proceed with SUSYQM and shape invariance. The function α_k needs to be determined from the actual solution of the related Schrödinger equation.

---

[3]Some properties of the Γ-function:

$$\Gamma(z + 1) = z\,\Gamma(z); \ \ \Gamma(z)\Gamma(1 - z) = \frac{\pi}{\sin(\pi z)}; \ \ \Gamma(z)^* = \Gamma(z^*); \ \ \text{Res}(\Gamma, -n) = \frac{(-1)^n}{n!}.$$

(2) Rosen-Morse II

The superpotential for this system is given by

$$A \tanh x + \frac{B}{A} \quad \text{with} \quad B < A^2 \ . \tag{16.16}$$

The asymptotic value of the superpotential W at $-\infty$ is $-A + B/A$. Hence, the Eq. (16.10) leads to

$$r^-_{\mathrm{RM\,II}}(k, A+1) = \left(\frac{-A + \frac{B}{A} - ik}{-A + \frac{B}{A} + ik} \right) r^-_{\mathrm{RM\,II}}(k, A) \tag{16.17}$$

$$= \left(\frac{A^2 + ikA - B}{A^2 - ikA - B} \right) r^-_{\mathrm{RM\,II}}(k, A)$$

$$= \frac{\left[A + \frac{ik}{2} + \frac{i}{2}\sqrt{4B - k^2} \right] \left[A + \frac{ik}{2} - \frac{i}{2}\sqrt{4B - k^2} \right]}{\left[A - \frac{ik}{2} + \frac{i}{2}\sqrt{4B - k^2} \right] \left[A - \frac{ik}{2} - \frac{i}{2}\sqrt{4B - k^2} \right]}$$

$$\times \ r^-_{\mathrm{RM\,II}}(k, A) \ .$$

Using arguments similar to that of the last example, we obtain

$$r^-_{\mathrm{RM\,II}}(k, A) =$$

$$\frac{\Gamma \left[A + \frac{ik}{2} + \frac{i}{2}\sqrt{4B - k^2} \right] \Gamma \left[A + \frac{ik}{2} - \frac{i}{2}\sqrt{4B - k^2} \right]}{\Gamma \left[A - \frac{ik}{2} + \frac{i}{2}\sqrt{4B - k^2} \right] \Gamma \left[A - \frac{ik}{2} - \frac{i}{2}\sqrt{4B - k^2} \right]} \ \alpha_k(A, B) \ .$$

$$\tag{16.18}$$

Again, this is as far as we can go with SUSYQM and shape invariance. We have extracted information about the structure of these coefficients that follow from SUSYQM. We still need to go to the Schrödinger equation to get complete information on reflection and transmission coefficients.

We will not further pursue one-dimensional scattering. We have been able to extract a substantial part of the structures of these reflection amplitudes, using shape invariance and SUSYQM. Now we will employ the machinery we have developed toward understanding three-dimensional scattering, a better model of what actually happens in the laboratory.

16.2 The Three Dimensional Case

Let us consider the scattering of plane waves from a spherically symmetric potential. The wave function in such a case can be written as a linear combination of an incident plane wave and a scattered outgoing spherical wave $(\frac{e^{ikr}}{r})$ modulated by a coefficient $f(k,\theta)$, known as the scattering amplitude. We choose our z-axis to be along the incident beam, hence the incident wave, far from the center of scattering, is described by the plane wave solution e^{ikz}. Since the system has azimuthal symmetry; i.e., any rotation about the z-axis leaves the system invariant, the scattered wave function is independent of the angle ϕ.

$$\psi(r,\theta) = \psi_{\text{inc.}} + \psi_{\text{sc.}}$$

$$\approx e^{ikz} + f(k,\theta)\frac{e^{ikr}}{r} \ . \tag{16.19}$$

This amplitude $f(k,\theta)$ contains all pertinent information about the potential. In this section, as in the one preceding, we will attempt to determine as much of the structure of $f(k,\theta)$ from SUSYQM and shape invariance as possible.

In three dimensional quantum mechanics, the momentum $\vec{p} \equiv (p_x, p_y, p_z)$ is represented by the operator $-i\hbar\vec{\nabla} \equiv \left(-i\hbar\frac{\partial}{\partial x}, -i\hbar\frac{\partial}{\partial y}, -i\hbar\frac{\partial}{\partial z}\right)$. Hence, the hamiltonian in three-dimensions is given by

$$H = -\frac{\hbar^2}{2m}\nabla^2 + V(\vec{r}) \equiv -\nabla^2 + V(\vec{r}) \ , \tag{16.20}$$

where we chose, as usual, $\hbar = 2m = 1$. If our potential is spherically symmetric; i.e. the potential is a function of the radial coordinate only $V(r, \theta, \phi) = V(r)$, it is convenient to express the hamiltonian and the Schrödinger equation in spherical coordinates. We obtain:

$$-\frac{1}{r^2}\left[\frac{\partial}{\partial r}\left(r^2\frac{\partial}{\partial r}\right) + \frac{1}{\sin\theta}\frac{\partial}{\partial\theta}\left(\sin\theta\frac{\partial}{\partial\theta}\right) + \frac{1}{\sin^2\theta}\frac{\partial^2}{\partial\phi^2} - V(r)\right]\psi = E\psi \ .$$

This can be written as

$$-\left(\frac{d^2}{dr^2} + \frac{2}{r}\frac{d}{dr} - \frac{L^2}{r^2} - V(r)\right)\psi = E\psi \ ,$$

where

$$L^2 = -\left(\frac{1}{\sin\theta}\frac{\partial}{\partial\theta}\left(\sin\theta\frac{\partial}{\partial\theta}\right) + \frac{1}{\sin^2\theta}\frac{\partial^2}{\partial\phi^2}\right)$$

is the square of the angular momentum operator $\vec{L}$ discussed in Chapter 8. Writing $\psi(r,\theta,\phi) = R_{E\ell}(r)Y_{\ell m}(\theta,\phi)$ and using

$$L^2 Y_{\ell m}(\theta,\phi) = \ell(\ell+1)Y_{\ell m}(\theta,\phi) \ ,$$

the Schrödinger equation becomes

$$-\left[\frac{d^2}{dr^2} + \frac{2}{r}\frac{d}{dr} - \frac{\ell(\ell+1)}{r^2} - V(r)\right]R_{E\ell}(r) = ER_{E\ell}(r) \ . \tag{16.21}$$

Let us first consider the case of free-particles; i.e. $V(r) = 0$. To that end, we rewrite the above equation after making two substitutions: $E = k^2$ and $\rho = kr$. This leads to

$$\left[\frac{d^2}{d\rho^2} + \frac{2}{\rho}\frac{d}{d\rho} + \left(1 - \frac{\ell(\ell+1)}{\rho^2}\right)\right] R_\ell(\rho) = 0 \ . \tag{16.22}$$

This equation is the spherical Bessel equation. Its two linearly independent solutions are spherical Bessel functions defined by

$$j_\ell(\rho) = (-1)^\ell \rho^\ell \left(\frac{1}{\rho}\frac{d}{d\rho}\right)^\ell \frac{\sin\rho}{\rho} \tag{16.23}$$

and spherical Neumann functions

$$n_\ell(\rho) = (-1)^{\ell+1} \rho^\ell \left(\frac{1}{\rho}\frac{d}{d\rho}\right)^\ell \frac{\cos\rho}{\rho} \ . \tag{16.24}$$

The functions j_ℓ's are finite at $\rho = 0$, and n_ℓ's diverge at the origin. Leading behavior of these functions near the origin can be derived and are given by

$$\lim_{\rho\to 0} j_\ell(\rho) \to \frac{2^\ell \, \ell!}{(2\ell+1)!}\rho^\ell \quad \text{and} \quad \lim_{\rho\to 0} n_\ell(\rho) \to -\frac{(2\ell)!}{2^\ell \, \ell!}\frac{1}{\rho^{\ell+1}} \ . \tag{16.25}$$

Problem 16.4. *Starting from Eqs. (16.23) and (16.24), derive the asymptotic forms of Eq. (16.25) for small ρ.*

For large values of ρ, the asymptotic forms are

$$\lim_{\rho\to\infty} j_\ell(\rho) \to \frac{\sin\left(\rho - \frac{\ell\pi}{2}\right)}{\rho} \quad \text{and} \quad \lim_{\rho\to\infty} n_\ell(\rho) \to -\frac{\cos\left(\rho - \frac{\ell\pi}{2}\right)}{\rho}. \tag{16.26}$$

Problem 16.5. *Derive the above asymptotic forms of spherical Bessel functions for large ρ given in Eq. (16.26).*

Let us now go back to Eq. (16.19). In order to relate the incoming plane waves to the outgoing spherical waves, we will express the incident plane waves as well as the outgoing waves in terms of spherical Bessel functions and Legendre polynomials. Let us first show that a plane waves can be written as

$$e^{ikz} \equiv e^{ikr\cos\theta} = \sum_\ell (2\ell+1)\, i^\ell j_\ell(kr) P_\ell(\cos\theta) \ . \tag{16.27}$$

In order to prove the above identity, let us recognize that free particle wave functions e^{ikz} are solutions of Eq. (16.22). Thus, we should be able to expand the free particle wave solution in terms of Bessel functions and spherical harmonics:

$$e^{ikz} = \sum_{\ell} \left[c_\ell \, j_\ell(kr) + d_\ell \, n_\ell(kr) \right] P_\ell(\cos\theta) \ . \tag{16.28}$$

Coefficients c_ℓ and d_ℓ are constants. Since a free particle wave function should have no singularity at the origin, all d_ℓ's must vanish. Now, the only coefficients we have to determine are the c_ℓ's. Note that in general we would have needed the spherical harmonics $Y_{\ell m}(\theta, \phi)$ to express the angular part of the wave function. However, since there is azimuthal symmetry, the free particle wave function cannot depend on angle ϕ. Thus, the only $Y_{\ell m}(\theta, \phi)$'s that can appear are those that have $m = 0$; i.e., $Y_{\ell 0}$. Using $Y_{\ell 0}(\theta, \phi) \propto P_\ell(\cos\theta)$, we arrived at Eq. (16.28). Vanishing of the coefficients d_ℓ leads to

$$e^{ik \, r \, \cos\theta} = \sum_{\ell} c_\ell j_\ell(kr) P_\ell(\cos\theta) \ . \tag{16.29}$$

Multiplying both sides by $P_n(\cos\theta)$ and then integrating over $\cos\theta$ and using the orthogonality of the Legendre polynomials, we obtain

$$\frac{2c_n j_n(kr)}{2n+1} = \int_{-1}^{1} P_n(\cos\theta) e^{ik \, r \, \cos\theta} d(\cos\theta)$$

$$= \int_{-1}^{1} e^{ikru} \, P_n(u) \, du$$

$$= \frac{e^{ikru}}{ikr} \, P_n(u) \Big|_{-1}^{1} - \int_{-1}^{1} \frac{e^{ikru}}{ikr} \frac{d}{du} P_n(u) \, du$$

$$= \left(\frac{e^{ikr}}{ikr} P_n(1) - \frac{e^{-ikr}}{ikr} P_n(-1) \right) - \int_{-1}^{1} \frac{e^{ikru}}{ikr} \frac{d}{du} P_n(u) \, du$$

$$= \left(\frac{e^{ikr}}{ikr} - \frac{e^{-ikr}}{ikr} (-1)^n \right) - \int_{-1}^{1} \frac{e^{ikru}}{ikr} \frac{d}{du} P_n(u) \, du$$

$$= e^{in\pi/2} \left(\frac{e^{i(kr - n\pi/2)}}{ikr} - \frac{e^{-i(kr - n\pi/2)}}{ikr} \right)$$
$$\quad - \int_{-1}^{1} \frac{e^{ikru}}{ikr} \frac{d}{du} P_n(u) \, du$$

$$= \frac{2i^n}{kr} \sin(kr - n\pi/2) - \int_{-1}^{1} \frac{e^{ikru}}{ikr} \frac{d}{du} P_n(u) \, du \ . \tag{16.30}$$

Now we will consider the asymptotic form of last equation. For large values of r, the expression on the left hand side of Eq. (16.30); i.e., $\frac{2c_n j_n(kr)}{2n+1}$ has the form $\frac{2c_n}{2n+1} \frac{1}{kr} \sin\left(kr - \frac{\ell\pi}{2}\right)$. The first term on the right appears to have a similar form. However, the integral $\int_{-1}^{1} \frac{e^{ikru}}{ikr} \frac{d}{du} P_n(u) du$ is of order of $\mathcal{O}\left(\frac{1}{r^2}\right)$ or smaller. Thus, from the large-r limit of Eq. (16.30), we see that we must have

$$\frac{2}{2\ell + 1} c_\ell = 2i^\ell ,$$

which implies $c_\ell = (2\ell + 1) i^\ell$. Hence, as claimed in Eq. (16.27), the plane wave can be expanded in terms of products of $j_\ell(kr)$ and $P_\ell(\cos\theta)$:

$$e^{ikr\cos\theta} = \sum_\ell (2\ell + 1) i^\ell j_\ell(kr) P_\ell(\cos\theta) . \tag{16.31}$$

This is known as the Rayleigh expansion, which for large values of r, goes to

$$e^{ik\cos\theta} \sim \sum_\ell (2\ell + 1) i^\ell \frac{\sin\left(kr - \frac{\ell\pi}{2}\right)}{kr} P_\ell(\cos\theta) . \tag{16.32}$$

Again using the completeness of the Legendre polynomials $\mathcal{P}_\ell(\cos\theta)$, we write

$$\cdot f(k, \theta) = \sum_\ell f_\ell(k) \mathcal{P}_\ell(\cos\theta) . \tag{16.33}$$

Substituting these in Eq. (16.19), we obtain

$$\psi(r, \theta) = \sum_\ell \left((2\ell + 1) i^\ell \frac{\sin\left(kr - \frac{\ell\pi}{2}\right)}{kr} + f_\ell(k) \frac{e^{ikr}}{r} \right) P_\ell(\cos\theta) . \tag{16.34}$$

To determine the exact form of the amplitude $f_\ell(k)$, let us compare it with a general solution of the Schrödinger equation. The scattering states, far from the center of scattering, are described by general free-particle states given by $A_\ell(k) j_\ell(kr) + B_\ell(k) n_\ell(kr)$. The asymptotic value is $\frac{1}{kr} \left[A_\ell(k) \sin\left(kr - \frac{\ell\pi}{2}\right) - B_\ell(k) \cos\left(kr - \frac{\ell\pi}{2}\right) \right]$, which can be written as

$$\psi(r, \theta) = \frac{C_\ell(k)}{kr} \sin\left(kr - \frac{\ell\pi}{2} + \delta_\ell(k)\right) , \tag{16.35}$$

where $\delta_\ell(k) = -\tan^{-1}\left(\frac{B_\ell(k)}{A_\ell(k)}\right)$ and $C_\ell(k) = \sqrt{A_\ell^2(k) + B_\ell^2(k)}$.

Problem 16.6. *Derive the above values for $\delta_\ell(k)$ and $C_\ell(k)$.*

For the free-particle case, since singular solutions $n_\ell(kr)$ are not allowed, $B_\ell(k) = 0$. This implies $\delta_\ell(k) = 0$. Thus, comparison with Eq. (16.32) reveals that the interaction introduces the phase shift $\delta_\ell(k)$ to the plane wave solution. Equating the right hand sides of Eqs. (16.34) and (16.35) and comparing coefficients of $e^{\pm ikr}$, we obtain

$$C_\ell e^{i\left(-\frac{\ell\pi}{2}+\delta_\ell\right)} = (2\ell+1) + 2ikf_\ell(k) \quad \text{and} \quad C_\ell e^{i\left(\frac{\ell\pi}{2}-\delta_\ell\right)} = (2\ell+1) e^{i\left(\frac{\ell\pi}{2}\right)} .$$

(16.36)

Solving for $f_\ell(k)$, we obtain

$$f_\ell(k) = \frac{(2\ell+1)}{2ik} \left[e^{2i\delta_\ell} - 1\right] = \frac{(2\ell+1)}{k} e^{i\delta_\ell} \sin \delta_\ell . \qquad (16.37)$$

Now we will try to relate the the phase shift $\delta_\ell(k)$'s, and hence "partial-wave decompositions" $f_\ell(k)$ of the scattering coefficients $f(k, \theta)$, to the interactions; i.e, to the potentials, using supersymmetric quantum mechanics. We will see the effect of SUSYQM and shape invariance on the scattering amplitude and thus the phase shift.

If we substitute $u = r R_{E\ell}(r)$, Eq. (16.21) can be viewed as an effective one-dimensional Schrödinger equation for the reduced variable u. It becomes

$$\left(-\frac{d^2}{dr^2} + \underbrace{V(r) + \frac{\ell(\ell+1)}{r^2}}_{V_{\text{effective}}}\right) u(r) = E u(r) . \qquad (16.38)$$

Equation (16.38) can be written in factorized form

$$\underbrace{\left[-\frac{d}{dr} + W\right]}_{A^+} \underbrace{\left[\frac{d}{dr} + W\right]}_{A^-} u(r) = k^2 u(r) . \qquad (16.39)$$

For example, the function $W(r)$ for an attractive Coulomb potential is

$$W(r, \ell) = \frac{e^2}{2(\ell+1)} - \frac{\ell+1}{r} .$$

Taking advantage of the machinery we developed in the previous section for the one-dimensional, Eqs. (16.3)–(16.5), we can write the following equation for the asymptotic solution for u:

$$\left[\frac{d}{dr} + W_\infty\right] \sin\left(kr - \frac{\ell\pi}{2} + \delta_\ell^{(-)}(k)\right) = N \sin\left(kr - \frac{\ell\pi}{2} + \delta_\ell^{(+)}(k)\right) ,$$

(16.40)

where N is a constant. Equating coefficients of $e^{\pm ikr}$ on both sides of the equation, we obtain

$$e^{2i\delta_\ell^{(+)}} = \left[\frac{ik + W_\infty}{ik - W_\infty}\right] e^{2i\delta_\ell^{(-)}} . \qquad (16.41)$$

So far, we have only used SUSYQM. If the potential is shape invariant, then we can relate phase shift $\delta_\ell^{(+)}$ with $\delta_{\ell+1}^{(-)}$. From shape invariance, we obtain

$$e^{2i\delta_{\ell+1}^{(-)}} = \left[\frac{ik + W_\infty}{ik - W_\infty}\right] e^{2i\delta_\ell^{(-)}} . \qquad (16.42)$$

This is a recursion relation for the phase shift $\delta_\ell^{(-)}$ for a shape invariant potential. Iterating this equation, we will be able to generate the phase shift, or at least as much as can be derived using SUSYQM. We will now apply the above to the Coulomb problems.

Problem 16.7. *Derive the condition given in Eq. (16.41).*

16.2.1 Coulomb Potential

The Coulomb superpotential is given by

$$W(r, \ell) = \frac{e^2}{2(\ell+1)} - \frac{\ell+1}{r} .$$

From Eq. (16.7), the reflection amplitude is then

$$e^{2\delta^{(-)}}(k, \ell+1) = \left(\frac{2ik(\ell+1) + e^2}{2ik(\ell+1) - e^2}\right) e^{2\delta^{(-)}}(k, \ell)$$

$$= \left(\frac{(\ell+1) - i\frac{e^2}{2k}}{(\ell+1) + i\frac{e^2}{2k}}\right) e^{2\delta^{(-)}}(k, \ell) .$$

Using $\Gamma(z+1) = z\,\Gamma(z)$, we see that this recursion relation leads to

$$S_{\text{Coul.}}^-(k, \ell) \equiv e^{2\delta^{(-)}}(k, \ell) = \frac{\Gamma\left(\ell - i\frac{e^2}{2k}\right)}{\Gamma\left(\ell + i\frac{e^2}{2k}\right)} \alpha_k(k) , \qquad (16.43)$$

where α_k is independent of ℓ. As noted earlier, we cannot go any farther using just SUSYQM and shape invariance. We would need the Schrödinger equation itself. A comparison with results in literature [4] shows that the factor α_k for Coulomb is a ℓ-independent phase factor

$$\alpha_k = \frac{\Gamma\left(1 + i\frac{e^2}{2k}\right)}{\Gamma\left(1 - i\frac{e^2}{2k}\right)} \ .$$

Problem 16.8. *For the General Pöschl-Teller potential given by $W = A\tanh r - B\coth r$ show that the scattering amplitude has the form*

$$e^{2i\delta_B^{(-)}} = (-)^B \frac{\Gamma(A - B + ik)}{\Gamma(A - B - ik)} \ \alpha_k \ .$$

Problem 16.9. *For the Eckart potential given by $W = -A\coth r + \frac{B}{A}$, show that the scattering amplitude has the form*

$$e^{2i\delta_A^{(-)}} = (-)^A \ \frac{\Gamma\left(A - \frac{ik}{2} + \sqrt{B^2 + \frac{k^2}{4}}\right)\Gamma\left(A - \frac{ik}{2} - \sqrt{B^2 + \frac{k^2}{4}}\right)}{\Gamma\left(A + \frac{ik}{2} + \sqrt{B^2 + \frac{k^2}{4}}\right)\Gamma\left(A + \frac{ik}{2} - \sqrt{B^2 + \frac{k^2}{4}}\right)} \ \alpha_k \ .$$

[4]F. Cooper, A. Khare and U.P. Sukhatme, "Scattering amplitudes for supersymmetric shape-invariant potentials by operator methods", *Jour. Phys. A: Math. Gen.* **21** (1988) L501-L508.

Chapter 17

Natanzon Potentials

In 1979, a general set of potentials were discovered for which the Schrödinger equation reduces to the hypergeometric equation. Since in fact all of the known translational shape invariant potentials do just that, they must be special cases of Natanzon potentials[1, 2].

We write down the differential forms for the operators A^+ and A^-. We discover that these lead to a potential, the Natanzon potential. We find that is a very general form of the underlying equation for all one- and two-term shape invariant potentials. The equation for the Natanzon potential $U(r)$ may be written as

$$U[z(r)] = \frac{-fz(1-z) + h_0(1-z) + h_1 z}{R(z)} - \frac{1}{2}\{z, r\} , \qquad (17.1)$$

f, h_0, and h_1 are constants. z is an implicit function of r, which is itself either the radial or linear coordinate for each shape-invariant superpotential. $R(z)$ is a general quadratic function $az^2 + bz + c$. $\{z, r\}$ is the Schwartzian derivative, defined by

$$\{z, r\} \equiv \frac{d^3 z/dr^3}{dz/dr} - \frac{3}{2}\left[\frac{d^2 z/dr^2}{dz/dr}\right]^2 . \qquad (17.2)$$

Thus, if we can obtain expressions for z, that will give us the functions of r which yield the superpotentials. Thence follows a procedure similar to that in Chapter 9: finding a second order differential equation involving the Natanzon potential, and obtaining the condition for removing the first order term, thus rendering it in the form of the the Schrödinger equation. From this, the general form of the Natanzon potential, subjected to two

[1]G. A. Natanzon, *Teor. Mat. Fiz.* **38**, 219 (1979).
[2]For details, see Gangopadhyaya A., Mallow JV., Sukhatme UP. "Translational shape invariance and the inherent potential algebra", *Physical Review* **A58** 4287-4292 (1998).

constraints on z, reduces to the set of translational shape invariant potentials. The constraints are associated respectively with Natanzon potential and shape invariance. The first is

$$\left(\frac{dz}{dr}\right) = \frac{2z(1-z)}{\sqrt{R(z)}}. \tag{17.3}$$

The second is

$$\left(\frac{dz}{dr}\right) = z^{1+\beta}(1-z)^{-\alpha-\beta}. \tag{17.4}$$

α and β are as-yet-undetermined constants. Once we have found them, we will be able to generate $z(r)$ and thence the superpotentials. Combining Eqs. (17.3) and (17.3), we obtain

$$z^{1+\beta}(1-z)^{-\alpha-\beta} = \frac{2z(1-z)}{\sqrt{R(z)}}. \tag{17.5}$$

After some computation, we find that there are only a finite number of values of α and β which satisfy Eq. (17.5).

Problem 17.1. *Find the allowed values of the sets (α, β).*

Integrating Eq. (17.4), we obtain $z(r)$. From this, we can use Eq. (17.3) to determine $R(z)$. Putting that back into the Natanzon potential, Eq. (17.1) will give us th superpotential.

Let us demonstrate the procedure for $\alpha = \beta = 0$. $dz/dr = z \rightarrow z = e^r$. From Eq. (17.5),

$$z = \frac{2z(1-z)}{\sqrt{R(z)}},$$

whence $R(z) = 4(1-z)^2$. The Natanzon potential is then

$$U[z(r)] = \frac{-fz(1-z) + h_0(1-z) + h_1 z}{4(1-z)^2} - \frac{1}{2}\{z, r\} \tag{17.6}$$

$$-\frac{1}{2}\{z, r\} = \frac{1}{4}.$$

Defining $A \equiv -f/4, B \equiv h_0/4, C \equiv h_1/4$, and substituting $z = e^r$,

$$U = A\frac{e^r}{1-e^r} + B\frac{1}{1-e^r} + C\frac{e^r}{(1-e^r)^2} + \frac{1}{4}.$$

Writing $e^r = e^{\frac{r}{2}} / e^{-\frac{r}{2}}$, some manipulation yields

$$U = -A\frac{\cosh\left(\frac{r}{2}\right) + \sinh\left(\frac{r}{2}\right)}{2\sinh\left(\frac{r}{2}\right)} - B\frac{\cosh\left(\frac{r}{2}\right) - \sinh\left(\frac{r}{2}\right)}{2\sinh\left(\frac{r}{2}\right)} + \frac{C}{4\sinh^2\left(\frac{r}{2}\right)} + \frac{1}{4}$$

$$= -\frac{(A+B)}{2}\coth\left(\frac{r}{2}\right) - \frac{(A+B)}{2} + \frac{C}{4}\operatorname{cosech}^2\left(\frac{r}{2}\right) + \frac{1}{4}$$

Since this must have the form $W^2 - \frac{dW}{dr}$, we can examine various possibilities, and eventually choose

$$W = \tilde{m}_1\coth\left(\frac{r}{2}\right) + \tilde{m}_2.$$

This generates U with the identifications

$$\tilde{m}_1\tilde{m}_2 = -\frac{A+B}{4}, \tilde{m}_1^2 + \tilde{m}_1 = \frac{C}{4}, \tilde{m}_1^2 + \tilde{m}_2^2 = -\frac{A+B}{2} + \frac{1}{4}.$$

This is the Eckart potential.

Problem 17.2. *Without doing any calculation, show that the case $\alpha = 0$, $\beta = -1$ yields the Eckart potential.*

Problem 17.3. *Obtain the Rosen-Morse potential from $\alpha = -1$, $\beta = 0$.*

The table below gives the values α, β, the functions $z(r)$ which they generate, and the concomitant superpotentials. Replacing the hyperbolic functions by their trigonometric analogs gives us the full set of two-term translational shape invariant potentials. It shows all allowed value of α, β and the superpotentials that they generate. Constants $\tilde{m}_1$ and $\tilde{m}_2$ are linear functions of of the Natanzon constants f, h_0, h_1.

α	β	$z(r)$	**Superpotential**	**Potential**
0	0	$z = e^r$	$\tilde{m}_1 \coth \frac{r}{2} + \tilde{m}_2$	Eckart
0	$-\frac{1}{2}$	$z = \sin^2 \frac{r}{2}$	$\tilde{m}_1 \operatorname{cosec} r + \tilde{m}_2 \cot r$	Gen. Pöschl-Teller trigonometric
0	-1	$z = 1 - e^r$	$\tilde{m}_1 \coth \frac{r}{2} + \tilde{m}_2$	Eckart
$-\frac{1}{2}$	0	$z = -\operatorname{sech}^2 \frac{r}{2}$	$\tilde{m}_1 \operatorname{cosech} r + \tilde{m}_2 \coth r$	Pöschl-Teller II
$-\frac{1}{2}$	$-\frac{1}{2}$	$z = \tanh^2 \frac{r}{2}$	$\tilde{m}_1 \tanh \frac{r}{2} + \tilde{m}_2 \coth \frac{r}{2}$	Gen. Pöschl-Teller
-1	0	$z = \frac{1 + \tanh \frac{r}{2}}{2}$	$\tilde{m}_1 \tanh \frac{r}{2} + \tilde{m}_2$	Rosen Morse

PART 3
Summary

Chapter 18

Summary

Throughout this book, we have seen the ways in which SUSYQM recasts quantum mechanics. It starts by expanding the ladder operator method for the harmonic oscillator. Like that method, it seeks to obtain eigenvalues in a simpler way than by directly solving the Schrödinger equation. It does so by introducing a superpotential, from which are generated partner potentials: apparently unrelated but having the same eigenvalue spectrum except for an additional ground state of one of them. SUSYQM reverses the standard paradigm of quantum mechanics. Standard quantum mechanics asks "For a given potential, what is its energy spectrum?" SUSYQM asks, "For a given energy spectrum, what potentials can have generated it?"

If this were the whole story, then SUSYQM would provide interesting insights into the cause of what were thought to be accidental eigenspectrum matches, such as that for the infinite well and the $\csc^2$ potentials. However, one would still have to know the eigenspectrum of one of the partners *a priori* from ordinary quantum mechanics. But this means that we would have in any case to solve the Schrödinger equation for that partner. Thus SUSYQM would appear to be of rather limited value. But this changes with the discovery of shape invariance. For certain partner potentials, one can be related to the other by a simple parameter change and an additive constant. This allows us to obtain the eigenvalues and eigenfunctions of a single potential by using raising operators on the ground states of its set of partners. Thus we have a method that, like the harmonic oscillator, can obtain the same information as one obtains from the Schrödinger equation: the eigenvalue spectrum without recourse to that equation, and the set of eigenfunctions from sequential application of raising operators to ground state eigenfunctions, the only ones obtained from a differential equation, and a simple one at that.

Shape invariant potentials are of several types, such as translational, multiplicative, and cyclic. We focused on the first of these, for which there are analytical solutions. We saw that there is little more than a handful of known one- and two-term translational shape invariant potentials. No others have been found (except those formed by isospectral transformation of these), but we cannot be sure that there are no others.

We discussed two methods for obtaining such potentials. The first is by a group-theoretical approach called "potential algebra." The second is by generating from the difference-differential defining equation for the superpotential an equivalent partial differential equation, and obtaining solutions using a variant of the method of separation of variables. Both methods generated all of the known translational shape invariant potentials, and only those. It is worth noting that the potential algebra method, like SUSYQM itself, reverses the paradigm from standard quantum mechanics, but extended beyond consideration of only the potential. Standard quantum mechanics asks, "Given an operator, what are its eigenvalues?" The potential algebra method asks "Given a set of eigenvalues, what operator could have generated them?"

We demonstrated that SUSYQM was not restricted to obtaining bound states, but was also capable of obtaining scattering states. We showed that the reflection and transmission amplitudes were the same for partner potentials: $r^+ = r^-, t^+ = t^-$. Then, invoking shape invariance, we were able to obtain expressions for separate reflection and transmission coefficients for each of the partner potentials.

We related SUSYQM to Quantum Hamilton-Jacobi theory (QHJ). We found a simple relation between the ground state quantum momentum function of QHJ and the superpotential. This led to a relation between partner momenta p^+ and p^-. Invoking shape invariance, we obtained recursion relations for each of the partners p^+ and p^- separately. The raising and lowering operations for shape invariant potentials were expressed as fractional linear transformations connecting quantum momenta.

We showed how SUSYQM could be used for investigation of singular potentials. We saw that the only case that needed investigation was that of the inverse square potential whose multiplying coefficient (strength) fell within a certain range.

We studied broken SUSY, for which the critical condition of setting the ground state energy to zero could not be met. We showed how a particular parameter shift for shape invariant potentials could transform broken SUSY into unbroken SUSY, thus rendering it solvable.

We showed how the paradigm for generating known functions such as the Hermite and Laguerre polynomials from the Schrödinger equation could be reversed: using these polynomials to generate the full solutions to their respective Schrödinger equations. SUSYQM proved capable of producing recursion relations for such polynomials.

We examined the Dirac equation for hydrogenic atoms, and found it to be "intrinsically supersymmetric"; i.e., the large and small parts of the Dirac eigenfunctions behaved like the eigenfunctions of partner potentials. This allowed us to apply the machinery of SUSYQM to obtain the hydrogenic fine structure spectrum. Doing so provided an explanation for the degeneracy of energy eigenvalues for states of different orbital angular momentum ℓ but the same total (orbital + spin) angular momentum j.

We studied the extension of the standard WKB method to its supersymmetric version, SWKB, for which the WKB approximation becomes exact if the potential $V = W^2 \pm \hbar W'$ is replaced by W^2. We noted that this odd behavior is still unexplained.

So, what is left to do?

- The list of known shape invariant potentials and their isospectral deformations must either be shown to be complete, or expanded. If the latter, it would be good to find an algorithm for doing so.

- Methods used for generating translational shape invariant potentials should be examined to see if they may be modified for other shape invariant potentials.

- Other polynomials, in addition to the Hermite and Laguerre which we have studied, should be examined to see how they may produce Schrödinger equations.

- The SWKB problem must be explained.

- The partial differential equation *ansatz* might be expanded to more than two terms, and perhaps generate hitherto unknown shape invariant potentials.

This is not an exhaustive list. As in all fields, the solving of one problem generates others. We hope that this text will be a stepping stone to further study of supersymmetric quantum mechanics, and perhaps even to its further development.

PART 4
Solutions to Problems

Chapter 19

Solutions to Problems

Chapter 1

Problem 1.1

Let us consider the time dependent Schrödinger equation for

$$H(\mathbf{r})\Psi(\mathbf{r},t) \equiv -\frac{\hbar^2}{2m}\nabla^2\Psi(\mathbf{r},t) + V(\mathbf{r})\ \Psi(\mathbf{r},t) = i\hbar\ \frac{\partial}{\partial t}\Psi(\mathbf{r},t) \qquad (19.1)$$

for a time-independent Hamiltonian[1] $H(\mathbf{r})$. In such cases, we can write $\Psi(\mathbf{r},t) = \psi(\mathbf{r})f(t)$. Substituting this trial solution, also known as an ansatz, into Eq. (1.1), we get

$$-\frac{\hbar^2}{2m}\ \frac{\nabla^2\psi(\mathbf{r})}{\psi(\mathbf{r})} + V(\mathbf{r}) = i\hbar\ \frac{df(t)/dt}{f(t)}\ . \qquad (19.2)$$

The above equation has to be valid for all values of $\mathbf{r}$ and t. Since the expression on the left-hand-side of Eq. ()19.2 is a function of $\mathbf{r}$, while the right-hand-side is function of t, the above equality can hold only if both sides are equal to a common constant, say E. Then, from

$$-\frac{\hbar^2}{2m}\ \frac{\nabla^2\psi(\mathbf{r})}{\psi(\mathbf{r})} + V(\mathbf{r}) = E\ ,$$

we get

$$\left(-\frac{\hbar^2}{2m}\ \nabla^2 + V(\mathbf{r})\right)\ \psi(\mathbf{r}) \equiv H\psi(\mathbf{r}) = E\psi(\mathbf{r})\ . \qquad (19.3)$$

---

[1] An isolated system that does not exchange energy with its surroundings is always time-independent.

Chapter 2

Problem 2.1

Let us consider the differential equation (2.22).

$$\frac{d^2\psi(x)}{dx^2} = \frac{1}{4}\,\omega^2\,x^2\psi(x)\ . \tag{19.4}$$

Substituting the function $\psi(x) = e^{-\omega x^2/4}$, we get

$$\frac{d^2\psi(x)}{dx^2} = \left(\frac{\omega^2 x^2}{4} - \frac{\omega}{2}\right)\psi(x)\ .$$

For large x; i.e., we get $x \gg \sqrt{2/\omega}$, we get

$$\frac{d^2\psi(x)}{dx^2} \approx \frac{1}{4}\,\omega^2 x^2 \psi(x)\ ,$$

and hence $\psi(x) = e^{-\omega x^2/4}$ is indeed an asymptotic solution of Eq. (2.22).

Problem 2.2

Let us substitute $\xi \equiv x\,\sqrt{\omega/2}$ and $K \equiv 2\,E/\omega$ into Eq. (19.5).

$$-\frac{d^2\psi(x)}{dx^2} + \frac{1}{4}\,\omega^2\,x^2\psi(x) = E\psi(x)\ . \tag{19.5}$$

Since $\frac{d}{dx} = \sqrt{\omega/2}\frac{d}{d\xi}$, the above equation can be written as

$$-\frac{\omega}{2}\,\frac{d^2\psi(x)}{d\xi^2} + \frac{\omega}{2}\,\xi^2\psi(x) = \frac{\omega}{2}\,K\,\psi(x)\ . \tag{19.6}$$

Now substituting $\psi(x) = C\,e^{-\xi^2/2}\,v(\xi)$, after some simplification we get the desired result

$$\frac{d^2v(\xi)}{d\xi^2} - 2\,\xi\,\frac{dv(\xi)}{d\xi} + (K-1)\,v(\xi) = 0\ . \tag{19.7}$$

Problem 2.3

To show that for $j \gg 1$, $v_j \approx C/(j/2)!$ obeys the recursion relation Eq. (2.28),

$$v_{j+2} \approx \frac{2}{j}v_j\ , \tag{19.8}$$

we compute approximate forms for v_j and v_{j+2}. They are given by $\frac{C}{(\frac{j}{2})!}$ and $\frac{C}{(\frac{j+2}{2})!}$ respectively. Dividing v_{j+2} by v_j, we get

$$\frac{v_{j+2}}{v_j} = \frac{(\frac{j}{2})!}{(\frac{j+2}{2})!} = \frac{(\frac{j}{2})!}{(\frac{j}{2}+1)!} = \frac{1}{(\frac{j}{2}+1)} \approx \frac{2}{j} . \tag{19.9}$$

Problem 2.4

From the Rodrigues formula

$$H_n(\xi) = (-1)^n e^{\xi^2} \frac{d^n}{d\xi^n} \left(e^{-\xi^2} \right) \tag{19.10}$$

we want to show that

$$H_0(\xi) = 1, \ H_1(\xi) = 2\xi, \ H_2(\xi) = 4\xi^2 - 2, \ H_3(\xi) = 8\xi^3 - 12\xi . \tag{19.11}$$

The Hermite polynomial $H_0(\xi)$ is given by $(-1)^0 e^{\xi^2} \left(e^{-\xi^2} \right)$, which is equal to 1.

The Hermite polynomial $H_1(\xi)$ is given by $(-1) e^{\xi^2} \frac{d}{d\xi} \left(e^{-\xi^2} \right)$. It is equal to $(-1) e^{\xi^2} \left(-2\xi e^{-\xi^2} \right) = 2\xi$.

The Hermite polynomial $H_2(\xi)$ is given by $(-1)^2 e^{\xi^2} \frac{d^2}{d\xi^2} \left(e^{-\xi^2} \right)$. It is equal to $e^{\xi^2} \left((4\xi^2 - 2)e^{-\xi^2} \right) = 4\xi^2 - 2$.

Similarly the other two Hermite polynomials are given by $H_3(\xi) = 8\xi^3 - 12\xi$ and $H_4 = 16\xi^4 - 48\xi^2 + 12$.

Problem 2.5

Let us start with the recursion relation:

$$H_{n+1}(\xi) - 2\xi H_n(\xi) + 2n H_{n-1}(\xi) = 0 .$$

Replacing ξ by $\sqrt{\frac{\omega}{2}} x$,

$$H_{n+1} - \sqrt{2\omega} x H_n + 2n H_{n-1} = 0 .$$

Using

$$\psi_n(x) = \left(\frac{\omega}{2} \right)^{1/4} \left[\sqrt{\pi} n! \, 2^n \right]^{-1/2} e^{-\omega x^2/4} H_n \left(x \sqrt{\frac{\omega}{2}} \right)$$

we get

$$\frac{\psi_{n+1}(x)}{\sqrt{(n+1)!\,2^{(n+1)}}} - \frac{\sqrt{2\omega}\,x\,\psi_n(x)}{\sqrt{n!\,2^n}} + \frac{2n\,\psi_{n-1}(x)}{\sqrt{(n-1)!\,2^{(n-1)}}} = 0 \ . \qquad (19.12)$$

Starting from the recursion relation

$$H'_n(\xi) = 2nH_{n-1}(\xi),$$

and following a similar procedure we get Eq. (2.38). Then, using Eqs. (19.12) and (2.38), we arrive at Eq. (2.39).

Problem 2.6

From Eq. (2.21), we have

$$-\frac{d^2\psi(x)}{dx^2} + \frac{1}{4}\omega^2 x^2\,\psi(x) = E\psi(x) \ .$$

Redefining the potential V

$$V \equiv \frac{1}{4}\omega^2 x^2 - E_0$$

where E_0 is the ground state energy, we get

$$-\frac{d^2\psi(x)}{dx^2} + \underbrace{\left[\frac{1}{4}\omega^2 x^2 - E_0\right]}_{V_{\text{SUSY}}}\psi(x) = (E - E_0)\psi(x) \ .$$

Now, defining $E_{\text{susy}} \equiv E - E_0$,

$$-\frac{d^2\psi(x)}{dx^2} + V_{\text{SUSY}}\psi(x) = E_{\text{SUSY},0}\psi(x) \ .$$

From the definition of $E_{\text{SUSY},0}$, we have $E_{\text{SUSY},0} = E_0 - E_0 = 0$. Shifting a potential cannot change any physics, since the choice of zero potential is arbitrary. Hence, the wave functions do not change.

Problem 2.7

The potential for the half-Harmonic Oscillator (HHO) is

$$V(x) = \begin{cases} \infty, & x < 0 \\ \frac{1}{4}\omega^2 x^2, & x \geq 0 \end{cases} \ . \qquad (19.13)$$

Let us denote the eigenstates of the full harmonic oscillator by ψ_n. The corresponding energies are given by

$$\left(n + \frac{1}{2}\right)\hbar\omega \qquad n = 0, 1, 2, \ldots \ .$$

Since the potential for the HHO, for $x > 0$, is identical to that of the full harmonic oscillator (HO), the set of eigenfunctions of HO form a possible list of eigenfunctions for HHO. However, we should eliminate all the even states of HO that do not vanish at the origin. Hence, the states that are eigenfunctions of HHO are: ψ_1, ψ_3, ψ_5, $\cdots$; i.e., odd states of the full harmonic oscillator states. The ground state of HHO ψ_1 is the first excited state of HO. Hence the ground state energy of HHO is $\frac{3}{2}\hbar\omega$. Energies of higher state are

$$\frac{3}{2}\hbar\omega, \ \frac{7}{2}\hbar\omega, \ \frac{11}{2}\hbar\omega, \ldots, \ \frac{2n+3}{2}\hbar\omega \qquad n = 0, \ 1, \ 2, \ \ldots$$

Problem 2.8

From Eq. (2.44), we have

$$\frac{d}{dr}\left(r^2 \frac{dR(r)}{dr}\right) + r^2 \left(\frac{e^2}{r} + E\right) R(r) = l(l+1) R(r) . \tag{19.14}$$

Let us substitute $R(r) = r^{-1} u(r)$. This gives $\frac{dR}{dr} = -r^{-2}u + r^{-1}\frac{du}{dr}$. Or equivalently, $r^2 \frac{dR}{dr} = -u + r\frac{du}{dr}$. Hence we have

$$\frac{d}{dr}\left(r^2 \frac{dR}{dr}\right) = \frac{d}{dr}\left(-u + r\frac{du}{dr}\right) = r\frac{d^2u}{dr^2} .$$

Substituting the above in Eq. (19.14), the radial equation finally reduces to

$$-\frac{d^2u}{dr^2} + \left[\frac{-e^2}{r} + \frac{l(l+1)}{r^2}\right] u = Eu . \tag{19.15}$$

Problem 2.9

From Eq. (2.50), we have

$$\rho^2 \frac{d^2u}{d\rho^2} \approx l(l+1)u. \tag{19.16}$$

Using the ansatz $u = \rho^m$, we get

$$m^2 - m - l(l+1) .$$

The two solutions are $\frac{1}{2}(1 \pm (2l+1))$; i.e, $m = -l$ or $m = l+1$. Hence the general solution of Eq. (2.50) is

$$u \sim C\rho^{l+1} + D\rho^{-l}.$$

Problem 2.10

The general solution of Eq. (19.16) is given by $u \sim C\rho^{l+1} + D\rho^{-l}$. The second term blows up near the origin. The behavior of the complete solution of Eq. (2.48) near the origin is then given by ρ^{l+1}. The behavior at large values of ρ has the form e^{ρ}. Let us substitute $u \sim \rho^{l+1}\,e^{-\rho}\,v(\rho)$ into Eq. (2.48). This gives

$$-2l\,v - 2\,v + 2l\,v' + 2\,v' + r\,v - 2\rho\,v' + \rho\,v'' - \rho\,v + \rho_0\,v = 0 .$$

Simplifying it, we get

$$\rho\frac{d^2v}{d\rho^2} + 2(l+1-\rho)\frac{dv}{d\rho} + [\rho_0 - 2(l+1)]v = 0 ,$$

which is same as in Eq. (2.51).

Problem 2.11

We want to show that for large j, the ansatz $v_{j+1} \approx \frac{2^{j+m}}{(j+m)!}\,v_0$ solves the equation

$$v_{j+1} \approx \frac{2}{j}\,v_j .$$

From the above ansatz, we have $v_{j+1} = \frac{2^{j+m}}{(j+m)!}\,v_0$ and $v_j = \frac{2^{j+m-1}}{(j+m-1)!}\,v_0$. Dividing one by the other, we get

$$\frac{v_{j+1}}{v_j} = \frac{2(j+m-1)!}{(j+m)!} = \frac{2}{(j+m)} .$$

For $j \gg 1$, the above reduces to

$$\frac{v_{j+1}}{v_j} \approx \frac{2}{j} .$$

Problem 2.12

The exact recursion relation is given by Eq. (2.53)

$$v_{j+1} = \frac{2(j+l+1)-\rho_0}{(j+1)(j+2l+2)}\,v_j , \quad \text{where } \rho_0 > 0 .$$

Since

$$\frac{2(j+l+1)-\rho_0}{(j+1)(j+2l+2)} < \frac{2(j+l+1)}{(j+1)(j+2(l+1))} < \frac{2(j+l+1)}{(j+1)(j+l+1)}$$

$$= \frac{2}{j+1} < \frac{2}{j} ,$$

Eq. (2.54) indeed over estimates Eq. (2.53).

Problem 2.13

To get a solution that does not diverge, the series of Eq. (2.53) must automatically truncate. This happens if there is a j_{max} such that

$$v_{j_{max}+1} = \frac{2(j_{max} + l + 1) - \rho_0}{(j_{max} + 1)(j_{max} + 2(l + 1))} = 0 .$$

This happens for $2(j_{max} + l + 1) = \rho_0 \equiv 2n$. Since $\rho_0 = \frac{e^2}{\sqrt{-E}}$, we get

$$E \rightarrow E_n = -\frac{e^4}{4n^2} .$$

Problem 2.14

For 2s and 2p states, the radial functions are given by

$$R_{2s} = \frac{1}{a\sqrt{a}} \left(1 - \frac{r}{2a}\right) e^{-r/2a} ; \quad \text{and} \quad R_{2p} = \frac{1}{2a^2\sqrt{6a}} r e^{-r/2a} .$$

Since the probability density is given by $P(r) = r^2 R^2(r)$, the point r_{max} where the probability would be maximum would be given by

$$\frac{dP}{dr} = 0 .$$

For the **2s** state,

$$\frac{d}{dr}\left(r^2 R^2\right) = \frac{d}{dr}\left(r^2 \left(1 - \frac{r}{2a}\right)^2 e^{-r/a}\right) = 0,$$

implies $r(r - 2a) + 4a^2 - 4ra - 2r + r^2 = 0$. Its solution is $r = 2a$.

Similarly for the **2p** state,

$$\frac{d}{dr} r^2 r^2 e^{-r/a} = \frac{d}{dr} r^4 e^{-r/a} = 0,$$

gives $4r^3 - \frac{r^4}{a} = 0$. Hence, Its solution is $r = 4a$.

Bohr assumed circular orbits. Thus $L = m\,vr = n\hbar$. The energy of bound states for circular orbits is $E = T + V = \frac{m\,v^2}{2} - \frac{e^2}{r}$. Combining this with the force law

$$\frac{m\,v^2}{r} = \frac{e^2}{r^2}$$

we obtain

$$E = -\frac{L^2}{2mr^2}$$

classically. But for a given n, $E_n = -\frac{n^2 \hbar}{2mr^2}$. This should be different for r_{2s} and r_{2p}, but it isn't. Thus the Bohr model with $E = E_n$ violates the classical model $E \sim \frac{1}{2mr^2}$.

Problem 2.15

The radial function for the **2s** state is given by $R_{2s} = \frac{1}{a\sqrt{a}} \left(1 - \frac{r}{2a}\right) e^{-r/2a}$. Hence the function $u = rR$ is given by

$$u_{2s} = r\, R_{2s} = \frac{r}{a\sqrt{a}} \left(1 - \frac{r}{2a}\right) e^{-r/2a}.$$

Keeping terms to first order in r, we expand $e^{-r/2a} \approx 1 - \frac{r}{2a}$. Thus, we get

$$u_{2s} \sim r\left(1 - \frac{r}{2a}\right)^2 \approx r\left(1 - \frac{r}{a}\right) = r - \frac{r^2}{a} \sim \frac{\rho}{\kappa} - \frac{\rho^2}{\kappa^2 a},$$

where $\rho \equiv \kappa r$. But $a = \frac{1}{n\kappa} = \frac{1}{2\kappa}$ for $n = 2$, gives

$$u_{2s} \sim \frac{\rho}{\kappa} - \frac{\rho^2}{\kappa^2} \cdot 2\kappa = \frac{1}{\kappa}(\rho - 2\rho^2).$$

This agrees with

$$u_{n_0} \sim \rho - n\rho^2.$$

Chapter 3

Problem 3.1

Using matrix representation for states and linear operators, we want to show that $\langle m | An \rangle = \langle A^+ m | n \rangle$.

$$\langle m | An \rangle = (m_1^* m_2^*)\, \begin{pmatrix} a_{11} & a_{12} \\ a_{21} & a_{22} \end{pmatrix} \begin{pmatrix} n_1 \\ n_2 \end{pmatrix} = (m_1^* \ m_2^*)\, \begin{pmatrix} a_{11} n_1 & a_{12} n_2 \\ a_{21} n_1 & a_{22} n_2 \end{pmatrix}$$

$$= m_1^*(a_{11} n_1 + a_{12} n_2) + m_2^*(a_{21} n_1 + a_{22} n_2). \qquad (19.17)$$

$$\langle A^+ m| \, n \rangle = (A^+ m)^{*T} \cdot \begin{pmatrix} n_1 \\ n_2 \end{pmatrix} = \left(\begin{pmatrix} a_{11}^* & a_{21}^* \\ a_{12}^* & a_{22}^* \end{pmatrix} (m_1 \ m_2) \right)^{*T} \cdot \begin{pmatrix} n_1 \\ n_2 \end{pmatrix}$$

$$= \begin{pmatrix} a_{11}^* m_1 + a_{21}^* m_2 \\ a_{12}^* m_1 + a_{22}^* m_2 \end{pmatrix}^{*T} \cdot \begin{pmatrix} n_1 \\ n_2 \end{pmatrix} \tag{19.18}$$

$$= \begin{pmatrix} a_{11} m_1^* + a_{21} m_2^* \\ a_{12} m_1^* + a_{22} m_2^* \end{pmatrix}^{T} \cdot \begin{pmatrix} n_1 \\ n_2 \end{pmatrix}$$

$$= \begin{pmatrix} a_{11} m_1^* + a_{21} m_2^* & a_{12} m_1^* + a_{22} m_2^* \end{pmatrix} \cdot \begin{pmatrix} n_1 \\ n_2 \end{pmatrix}$$

$$= a_{11} m_1^* n_1 + a_{21} m_2^* n_1 + a_{12} m_1^* n_2 + a_{22} m_2^* n_2$$

$$= m_1^*(a_{11} n_1 + a_{12} n_2) + m_2^*(a_{21} n_1 + a_{22} n_2) = \langle m|An \rangle \ . \tag{19.19}$$

Problem 3.2

Let us check the hermiticity of the operator $p = -i \frac{d}{dx}$. Recall that for a hermitian operator, $\int \psi^* \, A\phi \, dx = \int \left(A^\dagger \psi \right)^* \phi \, dx$.

Thus, we need to show that $\int \psi^* \left(-i \frac{d}{dx} \right) \phi \, dx = \int \left(-i \frac{d}{dx} \psi \right)^* \phi \, dx$.

Hence,

$$\int \psi^* \, p\phi \, dx = \int \psi^* \cdot \left(-i \frac{d\phi}{dx} \right) dx$$

$$= -i \int \left[\frac{d}{dx} (\psi^* \phi) - \left(\frac{d\psi^*}{dx} \right) \cdot \phi \right] dx$$

$$= -i \, (\psi^* \phi)|_{\text{Boundary}} + i \int \left(\frac{d\psi^*}{dx} \right) \cdot \phi \, dx$$

$$= -0 + \int \left(-i \frac{d\psi}{dx} \right)^* \cdot \phi \, dx$$

$$= \int (p\psi)^* \, \phi \, dx \ . \tag{19.20}$$

So p is hermitian. $\frac{d}{dx}$ is not hermitian. Integration by parts gives the wrong sign:

$$\int \psi^* \left(\frac{d\phi}{dx}\right) dx = \psi^* \phi|_{\text{Boundary}} - \int \left(\frac{d\psi}{dx}\right)^* \phi \, dx$$

$$= -\int \left(\frac{d\psi}{dx}\right)^* \phi \, dx .$$

Hence, the operator $\frac{d}{dx}$ is skew Hermitian:

$$\left(\frac{d}{dx}\right)^\dagger = -\frac{d}{dx} .$$

Problem 3.3

Assume operators A and B are hermitian $A^\dagger = A$ and $B^\dagger = B$. Then

$$(A + B)^\dagger = A^\dagger + B^\dagger = A^\dagger + B^\dagger = A + B .$$

So $A + B$, the sum of two hermitian operators A and B, is hermitian.

Now let us see whether the same holds for products as well. Consider the product AB of two hermitian operators A and B.

$$\langle \psi | AB \, \phi \rangle = \langle A^\dagger \psi | B \phi \rangle$$

$$= \langle B^\dagger A^\dagger \psi | \phi \rangle$$

$$= \langle BA \psi | \phi \rangle .$$

Hence, $(AB)^\dagger = B^\dagger A^\dagger = BA$, which is not necessarily equal to AB unless they commute.

Since an operator A commutes with itself, the n-th power of a hermitian operator A; i.e., $(A)^n$, is a hermitian operator.

Problem 3.4

For an infinitely deep potential of width π, the eigenfunctions are $|n\rangle \equiv \sqrt{\frac{2}{\pi}} \sin nx$. Then

$$\langle m | n \rangle = \frac{2}{\pi} \int_0^\pi \sin (mx) \sin (nx) dx = \delta_{nm} .$$

Thus, all eigenfunctions are orthonormal. Let us now check the normalization of the state $|\psi\rangle = \frac{1}{2} |1\rangle + \frac{\sqrt{3}}{2} |2\rangle$.

$$\langle \psi | \psi \rangle = \frac{1}{4} \langle 1 | 1 \rangle + \frac{3}{4} \langle 2 | 2 \rangle = \frac{1}{4} + \frac{3}{4} = 1 .$$

The state $\langle\psi|\psi\rangle$ is normalized. The average value of energy in this state is given by

$$\langle E \rangle = \langle\psi|H\psi\rangle = \left(\frac{1}{2}\langle 1| + \frac{\sqrt{3}}{2}\langle 2|\right)\left(\frac{1}{2}H\,|1\rangle + \frac{\sqrt{3}}{2}H\,|2\rangle\right)$$

$$= \frac{1}{4}\langle 1|H|1\rangle + \frac{3}{4}\langle 2|H|2\rangle + 0 + 0$$

$$= \frac{1}{4}\times 1 + \frac{3}{4}\times 4 = 3\frac{1}{4}\,. \tag{19.21}$$

We have used $\langle 1|2\rangle = 0$.

This does not correspond to any single measurement, since it is not an eigenvalue of any single state.

The probability of finding the particle with energy 4 for state $|\psi\rangle = C_1\,|1\rangle + C_2\,|2\rangle$ is given by $|C_2|^2$. For this case, the probability of $E = 4$ is $|C_2|^2 = \frac{3}{4}$.

Problem 3.5

We want to show that x commutes with potential $V(x)$. We start with showing that x commutes an arbitrary power of x: x^n. $[x, x^n] = [x\,x^n - x^n\,x] = [x^{n+1} - x^{n+1}] = 0$.

The commutator of x with $V(x)$ is given by

$$[V(x), x] = \left[\sum_{n=0}^{\infty}\frac{x^n}{n!}\frac{d^n V}{dx^n}\bigg|_{x_0}, x\right]$$

$$= \sum_{n=0}^{\infty}\frac{1}{n!}\frac{d^n V}{dx^n}\bigg|_{x_0}[x^n, x] = 0\,. \tag{19.22}$$

Problem 3.6

For any operator A, the evolution of its expectation value $\langle A\rangle$ is given by

$$\frac{d\langle A\rangle}{dt} = \langle[H, A]\rangle + \frac{\partial\langle A\rangle}{\partial t}\,.$$

For the momentum operator p, the rate of change of its expectation vale with time is given by

$$\frac{d\langle p\rangle}{dt} = i\left\langle\left[\frac{p^2}{2m}, p\right]\right\rangle + i\,\langle[V, \rho]\rangle\,. \tag{19.23}$$

Since p commutes with any power of itself, the first term on the right hand-side of the above equation is zero. Hence,

$$\frac{d\langle p \rangle}{dt} = i\,\langle [V, p] \rangle = i\,\langle Vp - pV \rangle$$

$$= \left\langle \psi \left| V \frac{d\psi}{dx} \right\rangle - \left\langle \psi \left| \frac{d}{dx}(V\psi) \right\rangle \right.$$

$$= \left\langle \psi | V \frac{d\psi}{dx} \right\rangle - \left\langle \psi \left| V \frac{d\psi}{dx} \right\rangle - \left\langle \psi \left| \frac{dV}{dx} \psi \right\rangle \right. \right.$$

$$= -\left\langle \frac{dV}{dx} \right\rangle . \tag{19.24}$$

Chapter 4

Problem 4.1

In problem 3.3 we showed that for a product of two operators A and B, $(AB)^\dagger = B^\dagger A^\dagger$. For operators $A^\pm$, we know that $(A^\pm)^\dagger = A^\mp$. Hence,

$$\left(A^+ A^- \right)^\dagger = \left(A^- \right)^\dagger \left(A^+ \right)^\dagger = A^+ A^- .$$

Thus, A^+A^- and A^-A^+ are self-adjoint operators. However, we will go beyond this proof and show it explicitly as well. The operators A^+, A^- are given by

$$A^- \equiv \frac{1}{\sqrt{\omega}} \left(\frac{d}{dx} + \frac{1}{2}\omega x \right) ; \quad A^+ \equiv \frac{1}{\sqrt{\omega}} \left(-\frac{d}{dx} + \frac{1}{2}\omega x \right) .$$

The number operator and its conjugate are given by

$$N \equiv A^+ A^- = \frac{1}{\omega} \left(-\frac{d}{dx} + \frac{1}{2}\omega x \right) \left(\frac{d}{dx} + \frac{1}{2}\omega x \right)$$

$$N^\dagger = \left(A^- \right)^\dagger \left(A^+ \right)^\dagger = \frac{1}{\omega} \left(\frac{d}{dx} + \frac{1}{2}\omega x \right)^\dagger \left(-\frac{d}{dx} + \frac{1}{2}\omega x \right)^\dagger . \tag{19.25}$$

However, as per results of problem 3.2, $\frac{d}{dx}$ is skew hermitian: $\left(\frac{d}{dx} \right)^\dagger = -\frac{d}{dx}$. Hence,

$$N^\dagger = \frac{1}{\omega} \left(-\frac{d}{dx} + \frac{1}{2}\omega x \right) \left(\frac{d}{dx} + \frac{1}{2}\omega x \right) = N .$$

N is hermitian.

Problem 4.2

$[A^-, A^+] = A^- A^+ - A^+ A^- = 1$; i.e., $M - N = 1$. One of the things we notice is that M and N commute. Let us denote their common eigenstates by $|n, m\rangle$. Then we have

$$N |n, m\rangle = n |n, m\rangle \quad \text{and} \quad M |n, m\rangle = m |n, n\rangle .$$

Thus $m = n + 1$.

Since, $N \equiv A^+ A^- = \frac{H}{\omega} - \frac{1}{2}$, we have

$$M \equiv A^- A^+ = \frac{H}{\omega} + \frac{1}{2} .$$

Hence, the energy E_m would be related to m by

$$E_m = \left(m - \frac{1}{2} \right) \omega \quad m = 1, 2, 3 \cdots .$$

Problem 4.3

From Eq. (4.21), we have

$$\left(\frac{d}{dx} + \frac{\omega x}{2} \right) \psi_0(x) = 0 .$$

Thus,

$$\int \frac{d\psi_0}{\psi_0} = - \int \frac{\omega x}{2} dx + C.$$

Integrating the above,

$$\ln \psi_0(x) = -\frac{\omega x^2}{4} + C; \quad \text{or} \quad \psi_0(x) = N e^{-\frac{\omega x^2}{4}} .$$

Normalizing $\psi_0(x)$; i.e., demanding that $|N|^2 \int_{-\infty}^{\infty} e^{-\frac{\omega x^2}{4}} dx = 1$,

$$N = \left(\frac{\pi}{2\omega} \right)^{1/4} .$$

And hence

$$\psi_0 = \left(\frac{\pi}{2\omega} \right)^{1/4} e^{-\omega x^2/4} .$$

Chapter 5

Problem 5.1

To show that the ground state eigenfunction of $\psi_0^-(x,a)$ can be written as $\psi_0^-(x,a) \sim e^{-\int_{x_0}^x W(x,a)\,dx}$, we start with the $\left(\frac{d}{dx} + W(x,a)\right)\psi^-(x,a) = 0$. We then have $\frac{1}{\psi^-}\frac{d\psi^-}{dx} = -W(x,a)$. Integrating this equation we get $\psi_0^-(x,a) \sim e^{-\int_{x_0}^x W(x,a)\,dx}$.

Problem 5.2

For the following superpotentials, we need to show that the ground state wave functions vanish at both end points of their domains.

(1) Finite domain: $W(x,\alpha) = \alpha \tan x$, where $\alpha > 0$. The domain for this system is $\left(-\frac{\pi}{2}, \frac{\pi}{2}\right)$. The ground state wave function is given by $\psi_0^-(x,a) \sim e^{-\int_0^x \alpha \tan x\,dx} \sim \cos^\alpha x$. Since $\cos x$ vanishes at $\pm\frac{\pi}{2}$, the wave function is well behaved at both ends.

(2) Semi-infinite domain: $W(x,\alpha,\beta) = \alpha - \beta \coth x$ with $\alpha > \beta$. The domain for this system is $(0,\infty)$. The ground state wave function is given by $\psi_0^-(x,a) \sim e^{-\int_{x_0}^x (\alpha - \beta \coth x)\,dx} \sim e^{-\alpha x}\sinh^\beta x$. Since $\sinh x$ vanishes at the origin and the exponential function vanishes at infinity, the wave function is well behaved at both ends.

(3) Infinite domain: $W(x,\gamma) = \gamma \tanh x$ with $\gamma > 0$. The domain for this system is $(-\infty,\infty)$. The ground state wave function is given by $\psi_0^-(x,a) \sim e^{-\int_{x_0}^x \gamma \tanh x\,dx} \sim \cosh^{-\gamma} x$. Since $\cosh x$ vanishes at $\mp\infty$, the wave function is well behaved at both ends.

Problem 5.3

For the three-dimensional harmonic oscillator $W(r,\omega,\ell) = \omega r - \frac{\ell}{r}$,

(1) Plot of $W(r,10,-1)$ as a function of r is given below (Figure 19.1).
(2) Let us now check the behavior of the ground state wave function $\psi_0^-(r,\omega,\ell) \sim \exp\left(-\int_0^\infty W(r,10,-1)\,dr\right)$. Integrating this expression, we get $\psi_0^-(r,\omega,\ell) = e^{-\left(\frac{1}{2}\omega r^2 + \ell \ln r\right)} = r^\ell e^{-\frac{1}{2}\omega r^2}$. For $\ell = -1$, the wave function blows up at the origin.

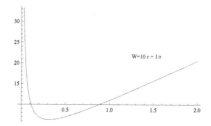

Fig. 19.1 Plot of $W(r, \omega, \ell) = \omega r - \frac{\ell}{r}$ as a function of r for $\omega = 10$ and $\ell = -1$.

Problem 5.4

To determine the normalization constants of ψ_n^+, we start with $\psi_n^+ = N\, A^-\, \psi_{n+1}^-$, where N is the normalization constant. Taking a scalar product of this state with itself, we get

$$\langle \psi_n^+ | \psi_n^+ \rangle = N^2 \left\langle \psi_{n+1}^- | A^+ A^- | \psi_{n+1}^- \right\rangle .$$

Since $A^+ A^- |\psi_{n+1}^-\rangle \equiv H_- |\psi_{n+1}^-\rangle = E_{n+1}^- |\psi_{n+1}^-\rangle$, we get $N = \frac{1}{E_{n+1}^-}$. We leave the other as an exercise.

Problem 5.5

$\psi_3^- (x, 1)$ is simply $\sin 3x$. Since $\psi_{n-1}^+(x, 1) \sim \left(\frac{d}{dx} - \cot x\right) \sin nx$, $\psi_2^+(x, 1)$ is given by $\left(\frac{d}{dx} - \cot x\right) \sin 3x = 3\cos 3x - \cot x \sin 3x = -8\sin^2(x)\cos(x)$. The wave functions are plotted in Figure 19.2.

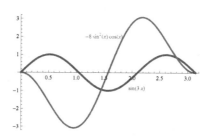

Fig. 19.2 Plots of $\psi_3^- (x, 1)$ and $\psi_2^+ (x, 1)$.

Problem 5.6

The graphs show that A^- acting on function $\sin 3x$ generates a function $-8\sin^2(x)\cos x$ that has one less node. Similarly, the operation of A^+ on function $\sin^2 x \cos x$ generates a function $-\sin 3x$ that has one additional node.

Problem 5.7

The superpotential is given by $W(x) = \Theta(x - x_0) - \frac{1}{2}$, where $\Theta(x - x_0)$ is a step function; i.e., $\Theta(x - x_0) = 1$ if $x > x_0$ and zero otherwise.

(1) The zero energy ground state wave function is given by

$$\psi_0^-(x, a) \sim e^{-\int_{x_0}^x \left(\Theta(x-x_0) - \frac{1}{2}\right) dx} \sim \begin{cases} e^{-\frac{1}{2}x} & \text{for } x > x_0 \\ e^{\frac{1}{2}x} & \text{for } x < x_0 \end{cases}.$$

This ground state vanishes sufficiently rapidly as $x \to \infty$, and hence is a normalizable state. This implies that supersymmetry is unbroken.

(2) A good representation of $\Theta(x - x_0) - \frac{1}{2}$ is given by $\lim_{\epsilon \to 0} \tanh\left(\frac{x-x_0}{\epsilon}\right)$. We have plotted partner potentials for $\epsilon = 0.01$. The potential $V_- = 1 - 99\,\text{sech}^2(100\,x)$ is given in Figure 19.3. The potential $V_+ = 1 + 99\,\text{sech}^2(100\,x)$ is given in Figure 19.4. As we see, the potential V_+ does not hold any boundstate. Thus, there are no state for the potential V_-

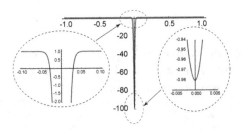

Fig. 19.3   Plots of $V_- = 1 - 99\,\text{sech}^2(100\,x)$.

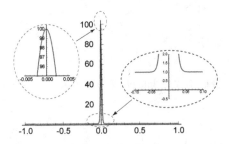

Fig. 19.4   Plots of $V_+ = 1 + 99\,\text{sech}^2(100\,x)$.

above the ground state. Thus, the attractive delta-function potential V_- holds only one bound state.

Chapter 6

Problem 6.1

From Chapter 6, the Coulomb superpotential is

$$W(r, \ell, e) = \frac{1}{2} \frac{e^2}{\ell + 1} - \frac{\ell + 1}{r}.$$

The resulting partner potentials are shape invariant since we have

$$V_+(r, \ell, e) = V_-(r, \ell + 1, e) + \underbrace{\frac{1}{4} \left(\frac{e^2}{\ell + 1} \right)^2}_{-g(\ell+1)} \underbrace{- \frac{1}{4} \left(\frac{e^2}{\ell + 2} \right)^2}_{-g(\ell+2)}. \qquad (19.26)$$

$$\underbrace{\phantom{\frac{1}{4} \left(\frac{e^2}{\ell + 1} \right)^2 - \frac{1}{4} \left(\frac{e^2}{\ell + 2} \right)^2}}_{R(a_0)}$$

The above equation implies

$$\psi_n^{(+)}(r, \ell, e) = \psi_n^{(-)}(r, \ell + 1, e).$$

The ground state of the hydrogen atom is

$$\psi_0^{(-)}(r, \ell, e) = r^{\ell+1} \exp \left(-\frac{1}{2} \frac{e^2}{\ell + 1} r \right). \qquad (19.27)$$

Now let us determine $\psi_1^{(-)}(r, \ell, e)$. Using SUSYQM $\psi_1^{(-)}(r, \ell, e) = A^+ \psi_0^{(+)}(r, \ell, e)$. From Eq. (19.26) we have $\psi_0^{(-)}(r, \ell + 1, e) = \psi_0^{(+)}(r, \ell, e)$. Hence,

$$\begin{aligned} \psi_1^{(-)}(r, \ell, e) &= A^+(r, \ell + 1, e) \, \psi_0^{(-)}(r, \ell + 1, e) \\ &= \left(-\frac{d}{dr} + W(r, \ell, e) \right) \psi_0^{(-)}(r, \ell + 1, e) \\ &= \left(-\frac{d}{dr} + \frac{1}{2} \frac{e^2}{\ell + 1} - \frac{\ell + 1}{r} \right) r^{\ell+2} \exp \left(-\frac{1}{2} \frac{e^2}{\ell + 2} r \right). \end{aligned}$$

$$(19.28)$$

Similarly

$$\psi_2^{(-)}(r, \ell, e) = A^+(r, \ell + 2, e) A^+(r, \ell + 1, e) \, r^{\ell+3} \exp \left(-\frac{1}{2} \frac{e^2}{\ell + 3} r \right).$$

$$(19.29)$$

Now let us check the conditions for the breaking of supersymmetry. From Eq. (19.27), we have $\left|\psi_0^{(-)}(r,\ell,e)\right|^2 = r^{2\ell+2}\exp\left(-\frac{e^2}{\ell+1}r\right)$. The integral of this function diverges near the origin if $2\ell + 3 < -1$, or if $\ell < -2$. This is the condition for the breaking of supersymmetry.

Problem 6.2

The superpotential for the Hulthen potential is

$$W(r,\ell) = \frac{a}{\ell} - \frac{\ell}{a}\left[\frac{1+e^{-\frac{r}{a}}}{1-e^{-\frac{r}{a}}}\right] = \frac{a}{\ell} - \frac{\ell}{2a}\coth\left(r/2a\right) , \qquad (19.30)$$

and corresponding coefficient $g(a_0)$ is

$$g(a_0) \equiv g(\ell) = -\left(\frac{\ell^2}{4a^2} + \frac{a^2}{\ell^2}\right).$$

The ground state $\sim e^{-\int W\,dx}$ is

$$\psi_0^{(-)} = e^{-\frac{a}{\ell}r}\left(\sinh\frac{r}{2a}\right)^{\frac{\ell}{2a}} .$$

The first excited state is then given by the following expression:

$$\left[\frac{a}{\ell} - \frac{\ell}{2a}\coth\left(r/2a\right)\right]e^{-\frac{a}{\ell+1}r}\left(\sinh\frac{r}{2a}\right)^{\frac{\ell+1}{2a}} .$$

The energies are

$$E_n = g(a_n) - g(a_0) ;$$
$$= \left(\frac{\ell+n^2}{4a^2} + \frac{a^2}{\ell+n^2}\right) - \left(\frac{\ell^2}{4a^2} + \frac{a^2}{\ell^2}\right) . \qquad (19.31)$$

Hence, $E_0 = 0$ and $E_1 = \left(\frac{\ell+1^2}{4a^2} + \frac{a^2}{\ell+1^2}\right) - \left(\frac{\ell^2}{4a^2} + \frac{a^2}{\ell^2}\right)$.

Problem 6.3

Since $W(x,a) = a\tanh x$, the partner potentials are given by

$$V_\pm(x,a) = a^2\tanh^2 x \pm a\,\text{sech}^2 x = a^2\tanh^2 x \pm a\left(1 - \tanh^2 x\right)$$
$$= a(a \mp 1)\,\text{sech}^2 x + a^2 . \qquad (19.32)$$

Thus, we have

$$V_+(x,a) = V_-(x,a-1) + \underbrace{a^2 - (a-1)^2}_{R(a_0)\equiv R(a)} .$$

Ground state: The energy for the ground state is zero and the corresponding eigenfunction is

$$\psi_0^{(-)}(x,a) = Ne^{-\int a\,\tanh x\,dx} = N\left(\cosh x\right)^{-a}.$$

First excited state: The energy for the first excited state is $R(a_0) = a^2 - (a-1)^2$, and the corresponding eigenfunction is given by

$$\psi_1^{(-)}(x,a) \sim A^+(x,a)\psi_0^{(-)}(x,a-1) = N\left(-\frac{d}{dx} + a\,\tanh x\right)\left(\cosh x\right)^{-(a-1)}.$$

We can proceed this way to determine eigenvalues and eigenfunctions of this system. The eigenvalues are given by $E_n = a^2 - (a-n)^2$ and the eigenfunctions are given by

$$\psi_n^{(-)}(x,a) \sim A^+(x,a)A^+(x,a-1)A^+(x,a-2)\cdots A^+(x,a-n+1)\psi_0^{(-)}(x,a-n).$$

These eigenfunctions are normalizable as long as $a > n$. Thus, the number of boundstates for this system is limited by the value of parameter a.

Problem 6.4

$$W(x,a,b) = a\tan x - b\cot x,$$

with $0 < x < \frac{\pi}{2}$ and $a > 0$, $b > 0$.

$$W(x,a,b) = a\tan x - b\cot x = a\,\frac{\sin x}{\cos x} + b\,\frac{\cos x}{\sin x}$$

$$= \frac{a\sin^2 x + b\cos^2 x}{\sin x \cos x}$$

$$= \frac{(b-a)\cos^2 x + a}{\sin x \cos x}$$

$$= (a+b)\operatorname{cosec} 2x + (b-a)\cot 2x$$

$$= c\operatorname{cosec} z + d\cot z$$

where $c = a+b$, $d = b-a$, and $z = 2x$.

Chapter 7

Problem 7.1

From Chapter 7, we have $(Q^-) = \begin{pmatrix} 0 & 0 \\ A^- & 0 \end{pmatrix}$. From this definition, we find $(Q^-)^2 = \begin{pmatrix} 0 & 0 \\ A^- & 0 \end{pmatrix}\begin{pmatrix} 0 & 0 \\ A^- & 0 \end{pmatrix} = \begin{pmatrix} 0 & 0 \\ 0 & 0 \end{pmatrix}$. Similarly, $(Q^+)^2 = \begin{pmatrix} 0 & A^+ \\ 0 & 0 \end{pmatrix}\begin{pmatrix} 0 & A^+ \\ 0 & 0 \end{pmatrix} = \begin{pmatrix} 0 & 0 \\ 0 & 0 \end{pmatrix}$.

Since the hamiltonian H is given by $\{Q^-, Q^+\} \equiv Q^-Q^+ + Q^+Q^-$, commutator of H with $Q^\mp$ can be computed as follows:

$$\begin{aligned}
[Q^-, H] &= [Q^-, Q^-Q^+ + Q^+Q^-] \\
&= [Q^- \cdot (Q^-Q^+ + Q^+Q^-) - (Q^-Q^+ + Q^+Q^-) \cdot Q^-] \\
&= [Q^-Q^+Q^- - Q^-Q^+Q^-] = 0 .
\end{aligned}$$

Similarly, $[Q^+, H] = 0$.

Problem 7.2

Since H is given by $\begin{pmatrix} H_- & 0 \\ 0 & H_+ \end{pmatrix} = \begin{pmatrix} A^+A^- & 0 \\ 0 & A^-A^+ \end{pmatrix}$, the action of H on $\begin{pmatrix} \psi_n^-(x) \\ 0 \end{pmatrix}$ leads to

$$\begin{pmatrix} A^+A^- & 0 \\ 0 & A^-A^+ \end{pmatrix}\begin{pmatrix} \psi_n^-(x) \\ 0 \end{pmatrix} = \begin{pmatrix} A^+A^-\psi_n^-(x) \\ 0 \end{pmatrix} = E_n^{(-)}\begin{pmatrix} \psi_n^-(x) \\ 0 \end{pmatrix} .$$

Similarly,

$$\begin{pmatrix} A^+A^- & 0 \\ 0 & A^-A^+ \end{pmatrix}\begin{pmatrix} 0 \\ \psi_{n-1}^+(x) \end{pmatrix} = \begin{pmatrix} 0 \\ A^-A^+\psi_{n-1}^+(x) \end{pmatrix} = E_{n-1}^{(+)}\begin{pmatrix} 0 \\ \psi_{n-1}^+(x) \end{pmatrix} .$$

From SUSYQM, we have $E_{n-1}^{(+)} = E_n^{(-)}$, hence $\begin{pmatrix} \psi_n^-(x) \\ 0 \end{pmatrix}$ and $\begin{pmatrix} 0 \\ \psi_{n-1}^+(x) \end{pmatrix}$ are degenerate eigenstates of the hamiltonian H.

Problem 7.3

Since $H = \{Q^-, Q^+\}$, the non-vanishing of $\langle 0|H|0 \rangle$ implies $\langle 0|Q^-Q^+ + Q^+Q^-|0 \rangle \neq 0$. However, both $\langle 0|Q^-Q^+|0 \rangle$ and $\langle 0|Q^+Q^-|0 \rangle$ are positive semi-definite since Q^- is the adjoint of Q^+; at least one of them has to be positive. Let us assume $\langle 0|Q^+Q^-|0\rangle \neq 0$. Since $\langle 0|Q^+Q^-|0\rangle = |Q^-|0\rangle|^2$, we get $|Q^-|0\rangle| \neq 0$. Thus, the vacuum is not invariant under supersymmetry.

Problem 7.4

$A^+A^-\phi^-(x) = 0$ implies $\int dx \, (\phi^-(x))^* A^+A^-\phi^-(x) = |A^-|\phi^-\rangle|^2 = 0$. This implies $|A^-|\phi^-\rangle| = 0$; hence $A^-|\phi^-\rangle = 0$, or in coordinate space $A^-\phi^-(x) = 0$.

Problem 7.5

$N_B - N_F = 0$ implies that we have broken supersymmetry. In such a case, the eigenstates of H_- and H_+ are in one-to-one correspondence. For every state of H_-, there is an eigenstate of H_+ with same eigenvalue. Hence,

$$
\begin{aligned}
Tr \; & \left((-)^F e^{-\beta H}\right) \\
&= (-1)^0 \left\langle \psi_0^{(-)}|e^{-\beta H}|\psi_0^{(-)}\right\rangle + (-1)^1 \left\langle \psi_0^{(+)}|e^{-\beta H}|\psi_0^{(+)}\right\rangle \\
&\quad + (-1)^0 \left\langle \psi_1^{(-)}|e^{-\beta H}|\psi_1^{(-)}\right\rangle + (-1)^1 \left\langle \psi_1^{(+)}|e^{-\beta H}|\psi_1^{(+)}\right\rangle + \cdots \\
&= \left\langle \psi_0^{(-)}|e^{-\beta E_0^{(-)}}|\psi_0^{(-)}\right\rangle - \left\langle \psi_0^{(+)}|e^{-\beta E_0^{(+)}}|\psi_0^{(+)}\right\rangle \\
&\quad + \left\langle \psi_1^{(-)}|e^{-\beta E_1^{(-)}}|\psi_1^{(-)}\right\rangle - \left\langle \psi_1^{(+)}|e^{-\beta E_1^{(+)}}|\psi_1^{(+)}\right\rangle + \cdots .
\end{aligned}
$$

Since, for broken supersymmetric cases, $E_n^{(-)} = E_n^{(+)}$, the sum in the above equation vanishes.

Problem 7.6

$N_B - N_F = 1$ implies that we have unbroken supersymmetry. In this case, the excited states of H_- and the eigenstates H_+ are in one-to-one correspondence. The ground state of H_-, corresponding to zero eigenvalue, has no partner among the states of H_+. Hence,

$$
Tr\left((-)^F e^{-\beta H}\right) = (-1)^0 \left\langle \psi_0^{(-)}|e^{-\beta H}|\psi_0^{(-)}\right\rangle = e^{-\beta E_0^{(-)}} = 1 .
$$

Chapter 8

Problem 8.1

The components of the angular momentum operator $\vec{L} = \vec{r} \times \vec{p}$ are:

$$L_x = yp_z - zp_y$$
$$L_y = zp_x - xp_z$$
$$L_z = xp_y - yp_x \ .$$

We write them in a compact notation by $L_i = x_j p_k - x_k p_j$, where i, j, k are in cyclic order.

In this problem, we will make extensive use of the identity

$$[AB,C] = A\,[B,C] + [A,C]\,B \ .$$

(1) Let us start with $[L_i, x_i]$. We assume i, j, k are a cyclic permutation of $1, 2, 3$.

$$[L_i, x_i] = [x_j p_k, x_i] - [x_k p_j, x_i]$$
$$= x_j [p_k, x_i] + [x_j, x_i] p_k - x_k [p_j, x_i] - [x_k, x_i] p_j \ .$$

Using $[x_i, p_j] = i\delta_{i,j} = 0$, and $[x_i, x_i] = 0$, we get $[L_i, x_i] = 0$. Similarly,

$$[L_i, x_j] = [x_j p_k - x_k p_j, x_j]$$
$$= [x_j p_k, x_j] - [x_k p_j, x_j]$$
$$= - [x_k p_j, x_j]$$
$$= -x_k [p_j, x_j] - [x_k, x_j] p_j = -x_k(-i) = ix_k \ .$$

Commutators $[L_i, p_i]$, $[L_i, p_j]$ are determined by very similar steps.

(2) Now let us determine commutators of the type: $[L_i, L_j]$.

$$[L_i, L_j] = [x_j p_k - x_k p_j, L_j]$$
$$= [x_j p_k, L_j] - [x_k p_j, L_j]$$
$$= x_j [p_k, L_j] + [x_j, L_j] p_k - x_k [p_j, L_j] - [x_k, L_j] p_j \ .$$

From the results of part (1):

$$1^{st} \text{ Term: } x_j [p_k, L_j] = -x_j [L_j, p_k] = -ix_j p_i$$
$$2^{nd} \text{ Term: } [x_j, L_j] p_k = - [L_j, x_j] p_k = 0$$
$$3^{rd} \text{ Term: } - x_k [p_j, L_j] = 0$$
$$4^{th} \text{ Term: } - [x_k, L_j] p_j = + [L_j, x_k] p_j = +ix_i p_j \ .$$

Putting all of these together, we get $[L_i, L_j] = i\,(x_i p_j - x_j p_i) = iL_k$. This can be written as

$$\left[\vec{L} \times \vec{L}\right] = i\vec{L} \ .$$

(3) Now we show that the square of the angular momentum operator L^2 commutes with each component L_i; i.e., $\left[L^2, L_i\right] = 0$.

$$
\begin{aligned}
\left[L^2, L_i\right] &= \left[L_i^2 + L_j^2 + L_k^2, L_i\right] \\
&= 0 + \left[L_j^2, L_i\right] + \left[L_k^2, L_i\right] \\
&= L_j\left[L_j, L_i\right] + \left[L_j, L_i\right]L_j + L_k\left[L_k, L_i\right] + \left[L_k, L_i\right]L_k \\
&= -\imath L_j L_k - \imath L_k L_j + \imath L_k L_j + \imath L_j L_k \\
&= 0 \ .
\end{aligned}
$$

Problem 8.2

(1) Up to a normalization constant, the action of the lowering operator $L_- = -e^{-\imath\phi}\left(\frac{\partial}{\partial\theta} - \imath\cot\theta\frac{\partial}{\partial\phi}\right)$ on $Y_2^2 \sim \sin^2\theta e^{2\imath\phi}$ gives:

$$
\begin{aligned}
L_- Y_2^2 &\sim -e^{-\imath\phi}\left(\frac{\partial}{\partial\theta} - \imath\cot\theta\frac{\partial}{\partial\phi}\right)\sin^2\theta e^{2\imath\phi} \\
&\sim -e^{-\imath\phi}\left[e^{2\imath\phi}2\sin\theta\cos\theta - \imath\cot\theta\sin^2\theta \cdot 2\imath e^{2\imath\phi}\right] \\
&\sim e^{-\imath\phi}\left[2\sin\theta\cos\theta + 2\sin\theta\cos\theta\right] \\
&\sim e^{\imath\theta} \cdot 4\sin\theta\cos\theta \\
&\sim e^{\imath\theta} \cdot \sin\theta\cos\theta \\
&\sim Y_2^1 \ .
\end{aligned}
$$

(2) We compute the actions of $L_\mp$ and L_3 on Y_2^2.

$$L_- Y_2^2 = 4Ne^{\imath\phi}\sin\theta\cos\theta \ ; \quad \text{where } N \text{ is the normalization constant } .$$

$$
\begin{aligned}
L_+ Y_2^2 &\sim -e^{-\imath\phi}\left(\frac{\partial}{\partial\theta} + \imath\cot\theta\frac{\partial}{\partial\phi}\right)\sin^2\theta e^{2\imath\phi} \\
&\sim -e^{-\imath\phi}\left[e^{2\imath\phi}2\sin\theta\cos\theta + \imath\cot\theta\sin^2\theta \cdot 2\imath e^{2\imath\phi}\right] \\
&\sim e^{-\imath\phi}\left[2\sin\theta\cos\theta - 2\sin\theta\cos\theta\right] \\
&= 0 \ .
\end{aligned}
$$

$$L_3 Y_2^2 = -\imath\frac{\partial}{\partial\phi}N\sin^2\theta e^{2\imath\phi} = 2\,Ne^{2\imath\phi}\sin^2\theta = 2\,Y_2^2 \ .$$

(3) Now we compute the action of L^2 on Y_2^2. For this, we use $L^2 = L_+ L_- + L_3^2 - L_3$.

$$L_+(L_-Y_2^2) = -e^{\imath\phi}\left(\frac{\partial}{\partial\theta} + \imath\cot\theta\frac{\partial}{\partial\phi}\right)4Ne^{\imath\phi}\sin\theta\cos\theta$$

$$= -4Ne^{\imath\phi}\left[e^{\imath\phi}\frac{\partial}{\partial\theta}\sin\theta\cos\theta + \imath\cot\theta\sin\theta\cos\theta\frac{\partial}{\partial\phi}e^{\imath\phi}\right]$$

$$= -4Ne^{\imath\phi}\left[e^{\imath\phi}(\cos^2\theta - \sin^2\theta) + \imath\cos^2\theta\imath\phi e^{\imath\phi}\right]$$

$$= -4Ne^{2\imath\phi}\left[\cos^2\theta - \sin^2\theta - \cos^2\theta\right]$$

$$= 4Ne^{2\imath\phi}\sin^2\theta \ .$$

$$(L_3)^2\,Y_2^2 = \imath\frac{\partial}{\partial\phi}\left(2Ne^{2\imath\phi}\sin^2\theta\right) = 4Ne^{2\imath\phi}\sin^2\theta \ .$$

Combining these operations, we get

$$\begin{aligned}\left(L_+L_- + L_3^2 - L_3\right)Y_2^2 &= 4Ne^{2\imath\phi}(-\sin^2\theta) + 4Ne^{2\imath\phi}\sin^2\theta - 2Ne^{2\imath\phi}\sin^2\theta\\ &= Ne^{2\imath\phi}\left[4\sin^2\theta + 4\sin^2\theta - 2\sin^2\theta\right]\\ &= 6Ne^{2\imath\phi}\sin^2\theta\\ &= 2\times 3Y_2^2\\ &= 2(2+1)Y_2^2 \quad\text{as expected.}\end{aligned}$$

Chapter 9

Problem 9.1

Apart from normalization constants, the eigenfunctions are

$$\psi_n(\xi) = N_n\,H_n(\xi)\,e^{-\frac{\xi^2}{2}} \ , \tag{19.33}$$

where $N_n = \left(\frac{1}{\pi}\right)^{\frac{1}{4}}\frac{1}{\sqrt{2^n n!}}$. We will now apply the SUSY operators $A^+(\xi,a_n) \equiv -\frac{d}{d\xi} + \xi$, and $A^-(\xi,a_n) \equiv \frac{d}{d\xi} + \xi$, on these eigenstates and use the following relationship we had derived from SUSYQM and shape invariance:

$$\psi_n^{(-)}(\xi,a_j) = \frac{1}{\sqrt{E_{n-1}^{(+)}}}\,A^+(\xi,a_j)\,\psi_{n-1}^{(-)}(\xi,a_{j+1}) \ .$$

Substituting for A^+ we obtain

$$N_n e^{-\frac{\xi^2}{2}}H_n(\xi) = \frac{1}{\sqrt{2n}}\left[-\frac{d}{d\xi} + \xi\right]N_{n-1}e^{-\frac{\xi^2}{2}}H_{n-1}(\xi) \ .$$

After some simplification we obtain the following recursion relation:

$$H_n(\xi) = \left[2\xi - \frac{d}{d\xi}\right] H_{n-1}(\xi) .$$

Similarly, using the identity that relates eigenfunctions of partner potentials via A^-, the following recursion relation can be generated:

$$2nH_{n-1}(\xi) = \left[\frac{d}{d\xi}\right] H_n(\xi).$$

Problem 9.2

We substitute $\theta = f(z)$ into

$$\frac{d^2 P_{\ell,m}}{d\theta^2} + \cot\theta \frac{dP_{\ell,m}}{d\theta} + \left[\ell(\ell+1) - \frac{m^2}{\sin^2\theta}\right] P_{\ell,m} = 0. \tag{19.34}$$

We use

$$\frac{dP_{\ell,m}}{d\theta} = \frac{dP_{\ell,m}}{dz}\frac{dz}{d\theta} = \frac{1}{f'}\frac{dP_{\ell,m}}{dz} ,$$

where $f' = \frac{d\theta}{dz}$. The second derivative is

$$\frac{d^2 P_{\ell,m}}{d\theta^2} = \frac{1}{f'}\frac{d}{dz}\left(\frac{1}{f'}\frac{dP_{\ell,m}}{dz}\right) = -\frac{f''}{f'^3}\frac{dP_{\ell,m}}{dz} + \frac{1}{f'^2}\frac{d^2 P_{\ell,m}}{dz^2}.$$

Substituting in Eq. (19.34), we obtain

$$\frac{1}{f'^2}\frac{d^2 P_{\ell,m}}{dz^2} + \left(\frac{\cot f}{f'} - \frac{f''}{f'^3}\right)\frac{dP_{\ell,m}}{dz} + \left[\ell(\ell+1) - \frac{m^2}{\sin^2 f}\right] P_{\ell,m} = 0 .$$

Multiplying the above equation by f'^2,

$$\frac{d^2 P_{\ell,m}}{dz^2} + \left[-\frac{f''}{f'} + f'\cot f\right]\frac{dP_{\ell,m}}{dz} + f'^2\left[\ell(\ell+1) - \frac{m^2}{\sin^2 f}\right] P_{\ell,m} = 0 .$$

Problem 9.3

Denoting differentiation with respect to x by prime, and suppressing sub- and superscripts,

$$xL'' + (\alpha + 1 - x)L' + nL = 0 .$$

We choose the ansatz $L \equiv Uf$, and demand that the coefficient of U' in the resulting differential equation vanish. $L' = Uf' + fU'$ and $L'' = Uf'' + 2f'U' + fU''$.

Then the differential equation becomes

$$x[fU'' + 2f'U' + Uf''] + (\alpha + 1 - x)[fU' + Uf'] + nUf = 0 .$$

Collecting terms in orders of U, the coefficient of U' is $2xf' + (\alpha + 1 - x)f$. $2xf' + (\alpha + 1 - x)f = 0$ has the solution

$$f = cx^{-\frac{(\alpha+1)}{2}} e^{\frac{x}{2}} .$$

Chapter 10

Problem 10.1

Given that

$$\hat{W}(x) = W(x) + \frac{d}{dx}\ln[I(x) + \lambda]$$

we want to show

$$\hat{V}_-(\lambda, x) = \hat{W}(x)^2 - \hat{W}(x)' = V_-(x) - 2\frac{d^2}{dx^2}\ln[I(x) + \lambda]$$

$$\hat{W}(x)^2 = \left(W(x) + \frac{d}{dx}\ln[I(x) + \lambda]\right)^2$$

$$= W(x)^2 + \underbrace{\left(\frac{d}{dx}\ln[I(x) + \lambda]\right)^2}_{f^2} + \underbrace{2W(x)\frac{d}{dx}\ln[I(x) + \lambda]}_{2W\,f}$$

$$= W(x)^2 - \frac{d^2}{dx^2}\ln[I(x) + \lambda]$$

where we used the results of Eq. (10.7). Thus, for $\hat{W}(x)^2 - \hat{W}(x)'$ we obtain

$$\hat{W}(x)^2 - \hat{W}(x)' = W(x)^2 - W(x)' - 2\frac{d^2}{dx^2}\ln[I(x) + \lambda]$$

$$\hat{V}_-(\lambda, x) = = V_-(x) - 2\frac{d^2}{dx^2}\ln[I(x) + \lambda] \ .$$

Problem 10.2

We want to show that these deformations do not change the reflection and transmission amplitudes. From Eq. (10.3), we have

$$R_-(k) = \frac{W_- + ik}{W_- - ik} R_+(k) \ , \qquad T_-(k) = \frac{W_+ - ik'}{W_- - ik} T_+(k) \ , \quad (19.35)$$

where $W_\pm$ are the asymptotic values of the superpotential. Let us compute the asymptotic values of the deformed superpotential $\hat{W}(x)$. Since the difference between the two superpotentials, $\hat{W}(x)$ and $W(x)$, is $\frac{d}{dx}\ln[I(x)+\lambda]$, we will focus on the asymptotic values of this function. From Eq. (10.11), this difference is given by

$$\frac{d}{dx}\ln[I(x) + \lambda] = \frac{\left[\psi_0^{(-)}(x)\right]^2}{\int_{-\infty}^x \left[\psi_0^{(-)}(t)\right]^2 dt + \lambda}$$

where $\psi_0^{(-)}(x)$ is the ground state wave function of the undeformed system. Since this must vanish for large distances, asymptotic values for this difference function are zero; i.e., $\hat{W}_\pm = W_\pm$. Hence, deformation does not alter the reflection and transmission amplitudes.

Problem 10.3

From Eq. (10.15), we have

$$\hat{\psi}_0^{(-)}(x) = \frac{\sqrt{\lambda(\lambda+1)}}{I(x)+\lambda}\,\psi_0^{(-)}(x)\ . \tag{19.36}$$

We assume that $\psi_0^{(-)}(x)$ is a normalized ground state wave function. We want to show that $\hat{\psi}_0^{(-)}(x)$ as given by Eq. (19.36) is normalized.

$$\int_{-\infty}^{\infty}\left|\hat{\psi}_0^{(-)}(x)\right|^2 dx = \int_{-\infty}^{\infty}\left|\frac{\sqrt{\lambda(\lambda+1)}}{I(x)+\lambda}\,\psi_0^{(-)}(x)\right|^2 dx$$

$$= \lambda(\lambda+1)\int_{-\infty}^{\infty}\frac{\left(\psi_0^{(-)}(x)\right)^2}{(I(x)+\lambda)^2}\,dx$$

$$= \lambda(\lambda+1)\int_{\lambda}^{\lambda+1}\frac{1}{\xi^2}\,d\xi$$

$$= \lambda(\lambda+1)\left(\frac{1}{\lambda}-\frac{1}{\lambda+1}\right)\ = 1\ .$$

In the above derivation, the variable ξ is defined by $\xi = (I(x)+\lambda)$. The limits of integration are obtained by noticing that $I(-\infty) = 0$, and $I(\infty) = 1$.

Problem 10.4

Without loss of generality we set $B = A$. We also choose the scale factor $\alpha = 1$. Then, the ground state of Morse potential has the form

$$\psi_0^{(-)}(x) = (2A)^A\,e^{-Ae^{-x}}\,e^{-Ax}\ . \tag{19.37}$$

This then leads to

$$I(x) \equiv \int_{-\infty}^{x}\left[\psi_0^{(-)}(t)\right]^2 dt = (2A)^{2A}\int_{-\infty}^{x}e^{-2Ae^{-t}}\left(e^{-2At}\right)dt\ . \tag{19.38}$$

The deformed ground state wave function is given by

$$\hat{\psi}_0^{(-)}(x) = \frac{\sqrt{\lambda(\lambda+1)}}{I(x)+\lambda}\,(2A)^A\,e^{-Ae^{-x}}\,e^{-Ax}\ .$$

For $\lambda = \infty$, the deformed wave function reduces to the undeformed $\psi_0^{(-)}(x)$. To plot, deformed wave function for other values of λ, let us set $A = 1/2$. Thus,

$$\psi_0^{(-)}(x) = e^{-\frac{1}{2}\left(x + e^{-x}\right)} \text{ , and } I(x) = \int_{-\infty}^{x} e^{-(t + e^{-t})} \, dt \ = \ e^{-e^{-x}}.$$

We see that the form of $I(x)$ obeys $I(-\infty) = 0$, and $I(\infty) = 1$. Wave function $\hat{\psi}_0^{(-)}(x)$ is now given by

$$\hat{\psi}_0^{(-)}(x) = \frac{\sqrt{\lambda(\lambda+1)}}{e^{-e^{-x}} + \lambda} \ e^{-\frac{1}{2}\left(x + e^{-x}\right)}.$$

For $\lambda = 0.2, 1.0$ and ∞ the deformed ground state wave functions are shown in Figure 19.5.

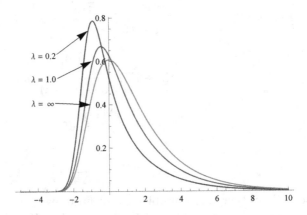

Fig. 19.5   Deformed ground state wave functions for $\lambda = 0.2, 1.0$ and ∞.

Problem 10.5

This problem is similar to the previous one for the Morse potential. For Coulomb, the ground state wave function is

$$\psi_0^{(-)}(r, q) = \frac{q^3}{\sqrt{2}} \ r \exp\left(-\frac{q^2}{2} \, r\right).$$

This then gives

$$I(r) \equiv \int_0^r \left[\psi_0^{(-)}(t)\right]^2 \, dt \ = \ \frac{1}{2}\left(2 - e^{-q^2 r}\left(q^2 r\left(q^2 r + 2\right) + 2\right)\right).$$

We see that, as expected, $I(0) = 0$ and $I(\infty) = 1$. Deformed wave function $\hat{\psi}_0^{(-)}(r,q)$ is then given by

$$\hat{\psi}_0^{(-)}(r,q) = \frac{\sqrt{\lambda(\lambda+1)}}{I(r)+\lambda} \, \psi_0^{(-)}(r,q)$$

$$= \frac{\sqrt{\lambda(\lambda+1)}}{\frac{1}{2}\left(2 - e^{-q^2 r}\left(q^2 r\left(q^2 r + 2\right) + 2\right)\right) + \lambda} \, \frac{q^3}{\sqrt{2}} \, r \exp\left(-\frac{q^2}{2}\,r\right) \ .$$

In Figure 19.6 we have drawn deformed ground state wave functions for $\lambda = 0.2, 1.0$ and ∞.

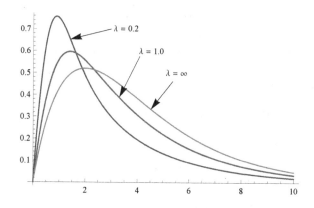

Fig. 19.6   Deformed ground state wave functions for $\lambda = 0.2, 1.0$ and ∞.

Chapter 11

Problem 11.1

The Morse superpotential is $W = A - Be^{-x}$. We want to check whether it satisfies

$$2W\frac{\partial W}{\partial a} - 2\frac{\partial W}{\partial x} + \frac{\partial g}{\partial a} = 0 \ .$$

Since the parameter A decreases by one under shape invariance related change of parameter while a_0 increases by one, we identify a_0 as $-A$.

$$a_0 \equiv -A \ ; \quad a_n = -A + n\hbar \ ; \quad \frac{\partial W}{\partial a} = -1 \ ; \quad \frac{\partial W}{\partial x} = Be^{-x} \ ; \quad \frac{\partial g}{\partial a} = -\frac{\partial g}{\partial A} \ .$$

Substituting into the above equation, we get

$$-2\left(A - Be^{-x}\right) - 2Be^{-x} = -2A = +\frac{\partial g}{\partial A} \ .$$

Hence

$$g(A) = \int 2A\, dA = -A^2 \ , \quad \text{and} \quad E_n = g(a_n) - g(a) = A^2 - (A - n)^2 \ .$$

Problem 11.2

The function $\chi(x)$ is

$$\chi = -\sqrt{\frac{H}{\frac{dA}{da}}} \tanh\left(\sqrt{H\frac{dA}{da}}\,(x - x_0)\right) \ .$$

Since χ is independent of a, $\rightarrow H \sim \frac{dA}{da}$ and $H \sim \left(\frac{dA}{da}\right)^{-1}$. This implies that H and $\frac{dA}{da}$ are constants. Replacement of H and $\frac{dA}{da}$ by γ and μ lead to the superpotential

$$W = (-\mu a + \beta)\sqrt{\frac{\gamma}{\mu}} \tanh\left[\sqrt{\gamma\mu}(x - x_0)\right]$$

where $H \equiv \gamma > 0$. Since

$$\tanh i\theta = \frac{\sinh i\theta}{\cos i\theta} = \frac{e^{i\theta} - e^{-i\theta}}{e^{i\theta} + e^{-i\theta}} = i\tan\theta \ ,$$

$\gamma < 0$ would imply that we have a trigonometric potential

$$W(x, a) = A\tan x \ .$$

Problem 11.3

For $A \equiv$ pure constant, $\frac{dA}{da} = 0$ and $\chi^2 \frac{dA}{da} = 0$. Then from Eq. (11.9), we find that $\frac{\partial\chi}{\partial x} = \mu$, a pure constant. This implies $\chi = \mu x + \beta$.

$$W = A\chi = A\left(\mu x + \beta\right) = \frac{1}{2}\omega\left(x - x_0\right) \ .$$

Problem 11.4

Let us consider a two-term superpotential

$$W(x, a) = A_1(a)\chi_1(x) + A_2(a)\chi_2(x) \ .$$

Assume

$$A_2(a) = qA_1(a) \ .$$

Then

$$W(x,a) = A_1(a)\chi_1(x) + qA_1(a)\chi_2(x) = A_1(a)\left(\chi_1(x) + q\chi_2(x)\right) .$$

This is of form $A_1(a)\chi_3(x)$ where $\chi_3 \equiv \chi_1(x) + q\chi_2(x)$.

Problem 11.5

Let us start with constraints given in Eq. (11.16)

$$\mathcal{F}_1(a_1)G_1(x) + \mathcal{F}_2(a_1)G_2(x) = 1$$
$$\mathcal{F}_1(a_2)G_1(x) + \mathcal{F}_2(a_2)G_2(x) = 1 .$$

For these equations to have the undesirable unique solution of G_1 and G_2 in terms of coefficients $\mathcal{F}_1$ and $\mathcal{F}_2$, the determinant of the coefficients must be non-zero. Therefore, to obtain the desired result: G's independent of $\mathcal{F}$'s, the determinant needs to be zero.

$$\begin{vmatrix} \mathcal{F}_1(a_1) & \mathcal{F}_2(a_1) \\ \mathcal{F}_1(a_2) & \mathcal{F}_2(a_2) \end{vmatrix} = 0$$

$$\Rightarrow \mathcal{F}_1(a_1)\mathcal{F}_2(a_2) = \mathcal{F}_2(a_1)\mathcal{F}_1(a_2) ;$$

i.e.,

$$\frac{\mathcal{F}_2(a_1)}{\mathcal{F}_1(a_1)} = \frac{\mathcal{F}_2(a_2)}{\mathcal{F}_1(a_2)} .$$

Problem 11.6

Let us consider the combination

$$\{1,2\} + \{3\} + \{4\} + (\{5\} \neq 0).$$

From $\{5\} \neq 0$ we have $\chi_1 \sim \frac{1}{\chi_2}$. Thus $\chi_2 = $ constant is not a solution. From $\{3\}$ and $\{4\}$, we find that χ_2 must be a linear function of x and $\mathcal{A}_2$ must be a constant. Thus the only solution for χ_1 that is compatible with $\chi_1 \sim 1/\chi_2$ and linearity of χ_2 is $\chi_1 = -\frac{\alpha}{x}$. The superpotential generated is $W(x,a) = -\frac{\alpha(\alpha a + \beta)}{x} + \gamma x$. With the identification of $x \to r$, $\alpha(\alpha a + \beta) \to \ell + 1$, and $\gamma \to \frac{1}{2}\omega$, we get the superpotential of the *three-dimensional harmonic oscillator*:

$$W(r,a) = \frac{1}{2}\omega r - \frac{l+1}{r} .$$

Chapter 12

Problem 12.1

Near the origin, Rosen-Morse, Eckart and Pöschl-Teller potentials are described by $-A/x$, $-A/r$ and $-(A+B)/r$ respectively. The behavior of the corresponding potentials (V_-) are

$$\text{Rosen-Morse}: \; V_-(x, A, B) \approx \frac{A(A+1)}{x^2}$$

$$\text{Eckart}: \; V_-(x, A, B) \approx \frac{A(A+1)}{r^2}$$

$$\text{Pöschl-Teller}: \; V_-(x, A, B) \approx \frac{(A+B)(A+B+1)}{r^2} \; .$$

Hence, corresponding α's are $A(A+1)$, $A(A+1)$ and $(A+B)(A+B+1)$ respectively.

Problem 12.2

The normalization condition for wave function ψ is given by

$$\int_0^{r_0} |\psi|^2 \, dr = 1.$$

Let us assume that the behavior of ψ near the origin is of the form r^m. Then, the above integral yields

$$\frac{r^{2m+1}}{2m+1} \Big|_0^{r_0} \; .$$

For $m > -\frac{1}{2}$, $2m + 1$ is a positive number, hence the integral does not blow up near the origin. $m = -\frac{1}{2}$ leads to a logarithmic divergence near the origin.

Problem 12.3

We want to solve the following approximate differential equation:

$$-\frac{d^2\psi}{dr^2} + \frac{\alpha}{r^2}\psi \approx 0 \; . \tag{19.39}$$

Multiplying Eq. (19.39) by r^2, we obtain

$$-r^2\frac{d^2\psi}{dr^2} + \alpha\,\psi \approx 0 \; .$$

Let us substitute the ansatz: $\psi = r^m$. This leads to $m^2 - m - \alpha = 0$. Its solutions are

$$m = \frac{1}{2}\left(1 \pm \sqrt{1 + 4\alpha}\right) .$$

Thus, the approximate solution for ψ near the origin is a linear combination of two possible solution:

$$\psi \approx c_1\, r^{\left(\frac{1}{2} - \sqrt{\alpha + \frac{1}{4}}\right)} + c_2\, r^{\left(\frac{1}{2} + \sqrt{\alpha + \frac{1}{4}}\right)} . \tag{19.40}$$

Problem 12.4

For $-\frac{1}{4} \leq \alpha < \frac{3}{4}$, we want to check the behavior of both terms of Eq. (19.40) near the origin. Let us consider the exponent $\left(\frac{1}{2} - \sqrt{\alpha + \frac{1}{4}}\right)$ at the two extreme values of α. For $-\frac{1}{4} \leq \alpha < \frac{3}{4}$, the exponent $-\frac{1}{2} < \frac{1}{2} - \sqrt{\alpha + \frac{1}{4}} \leq \frac{1}{2}$ and the exponent $\frac{1}{2} \leq \frac{1}{2} + \sqrt{\alpha + \frac{1}{4}} < \frac{3}{2}$. Thus, neither of the terms Eq. (19.40) diverge near the origin.

Problem 12.5

From Eq. (12.5), we have

$$-\frac{d^2\psi}{dr^2} + \frac{\alpha}{r_0{}^2}\psi = 0 . \tag{19.41}$$

It is a linear homogeneous differential equation with constant coefficients. For such equations, the trial solution used is $\psi(r) = e^{mr}$. Substituting this ansatz in the above equation we obtain

$$-m^2 e^{mr} + \frac{\alpha}{r_0{}^2} e^{mr} = 0.$$

Constant m is then given by $\pm\frac{\sqrt{\alpha}}{r_0}$. Hence, a general solution of the differential equation is given by

$$\psi(r) = Ce^{\frac{\sqrt{\alpha}}{r_0}r} + De^{-\frac{\sqrt{\alpha}}{r_0}r}$$

$$= \frac{1}{2}C\left(e^{\frac{\sqrt{\alpha}}{r_0}r} + e^{-\frac{\sqrt{\alpha}}{r_0}r} + e^{\frac{\sqrt{\alpha}}{r_0}r} - e^{-\frac{\sqrt{\alpha}}{r_0}r}\right)$$

$$+ \frac{1}{2}D\left(e^{\frac{\sqrt{\alpha}}{r_0}r} + e^{-\frac{\sqrt{\alpha}}{r_0}r} - e^{\frac{\sqrt{\alpha}}{r_0}r} + e^{-\frac{\sqrt{\alpha}}{r_0}r}\right)$$

$$= C\left(\cosh\frac{\sqrt{\alpha}}{r_0}r + \sinh\frac{\sqrt{\alpha}}{r_0}r\right) + D\left(\cosh\frac{\sqrt{\alpha}}{r_0}r - \sinh\frac{\sqrt{\alpha}}{r_0}r\right)$$

$$= (C+D)\cosh\frac{\sqrt{\alpha}}{r_0}r + (C-D)\sinh\frac{\sqrt{\alpha}}{r_0}r$$

$$= A\cosh\frac{\sqrt{\alpha}}{r_0}r + B\sinh\frac{\sqrt{\alpha}}{r_0}r. \tag{19.42}$$

Problem 12.6

Since we will be taking a limit $r_0 \to 0$, the Eq. (12.3) still governs the wave function just outside the well of radius r_0. We have derived its solution in problem 12.3, and is given by the linear combination

$$\psi \approx c_1 r^{\left(\frac{1}{2}-\sqrt{\alpha+\frac{1}{4}}\right)} + c_2 r^{\left(\frac{1}{2}+\sqrt{\alpha+\frac{1}{4}}\right)}. \tag{19.43}$$

Problem 12.7

Since $W(r,\alpha) = -\frac{\left(1+\sqrt{1+4\alpha}\right)}{2r}$, the partner potentials are

$$V_-(r,\alpha) = \left(-\frac{\left(1+\sqrt{1+4\alpha}\right)}{2r}\right)^2 + \frac{d}{dr}\frac{\left(1+\sqrt{1+4\alpha}\right)}{2r}$$

$$= \frac{\left(1+1+4\alpha+2\sqrt{1+4\alpha}\right)}{4r^2} - \frac{\left(1+\sqrt{1+4\alpha}\right)}{2r^2}$$

$$= \frac{\alpha}{r^2}.$$

Similarly,

$$V_+(r,\alpha) = \left(-\frac{\left(1+\sqrt{1+4\alpha}\right)}{2r}\right)^2 - \frac{d}{dr}\frac{\left(1+\sqrt{1+4\alpha}\right)}{2r}$$

$$= \frac{\left(1+1+4\alpha+2\sqrt{1+4\alpha}\right)}{4r^2} + \frac{\left(1+\sqrt{1+4\alpha}\right)}{2r^2}$$

$$= \frac{\left(1+\sqrt{1+4\alpha}\right)\left(3+\sqrt{1+4\alpha}\right)}{4r^2}.$$

Problem 12.8

From Eqs. (12.24) and (12.25), we have

$$W^2(x, a_0) + W'(x, a_0) = W^2(x, a_1) - W'(x, a_1) + \omega_0 , \qquad (19.44)$$
$$W^2(x, a_1) + W'(x, a_1) = W^2(x, a_0) - W'(x, a_0) + \omega_1 .$$

Adding these equations we obtain

$$W'(x, a_0) + W'(x, a_1) = \frac{1}{2}(\omega_0 + \omega_1) \equiv \omega .$$

I.e.,

$$W(x, a_0) + W(x, a_1) = \omega x ; \quad \text{or} \quad W(x, a_1) = \omega x - W(x, a_0) .$$

Substituting $W(x, a_1)$ in Eq. (19.44), we obtain

$$W^2(x, a_0) + W'(x, a_0) = (\omega x - W(x, a_0))^2 - (\omega - W'(x, a_0)) + \omega_0 .$$

This yields

$$2\omega x W(x, a_0) = \omega^2 x^2 - \omega + \omega_0 .$$

Dividing this equation by $2\omega x$, we obtain

$$W(x, a_0) = \frac{1}{2}\omega x + \frac{1}{2}\frac{\Omega}{\omega}\frac{1}{x} , \qquad (19.45)$$

where $\omega = \frac{1}{2}(\omega_0 + \omega_1)$ and $\Omega = \frac{1}{2}(\omega_0 - \omega_1)$.

Problem 12.9

We want to show that for all values of ω_0 and ω_1, the quantity $\frac{1}{4}\frac{\Omega(\Omega+2\omega)}{\omega^2}$ obeys the constraint

$$-\frac{1}{4} \le \frac{1}{4}\frac{\Omega(\Omega + 2\omega)}{\omega^2} < \frac{3}{4} ,$$

where $\omega = \frac{1}{2}(\omega_0 + \omega_1)$ and $\Omega = \frac{1}{2}(\omega_0 - \omega_1)$. Please note that ω_0 and ω_1, both need to be positive. This implies, $-\infty < \Omega < \infty$ and $0 < \omega < \infty$. The fraction $\frac{\Omega(\Omega+2\omega)}{\omega^2}$ can be written as

$$\frac{\Omega^2}{\omega^2} + 2\frac{\Omega}{\omega} = \left(\frac{\Omega}{\omega} + 1\right)^2 - 1 . \qquad (19.46)$$

Since $-1 \le \frac{\Omega}{\omega} \le 1$, the quantity $\frac{\Omega^2}{\omega^2} + 2\frac{\Omega}{\omega}$ obeys

$$-1 \le \frac{\Omega^2}{\omega^2} + 2\frac{\Omega}{\omega} \le 3 .$$

Thus,

$$-\frac{1}{4} \le \frac{1}{4}\frac{\Omega(\Omega+2\,\omega)}{\omega^2} < \frac{3}{4} .$$

Problem 12.10

For a homogeneous linear differential equation of the type

$$\frac{d^2\psi}{dr^2} + p(r)\frac{d\psi}{dr} + q(r)\psi = 0 ;$$

if one solution is $f(r)$, then the other solution can be obtained[2] by setting $\psi(r) = f(r)v(r)$. The substitution of $\psi(r) = f(r)v(r)$ reduces the equation to a first order in $\frac{dv}{dr}$. This is then integrated to yield

$$v(r) = \int \frac{e^{-\int p(r)dr}}{[f(r)]^2} \, dr . \qquad (19.47)$$

In our case the differential equation is

$$-\frac{d^2\psi}{dr^2} + \frac{\alpha}{r^2} \, \psi = 0 .$$

Two linearly independent solutions for ψ are

$$r^{\left(\frac{1}{2}-\sqrt{\alpha+\frac{1}{4}}\right)} \quad \text{and} \quad r^{\left(\frac{1}{2}+\sqrt{\alpha+\frac{1}{4}}\right)} .$$

For $\alpha = -\frac{1}{4}$, both solutions are identical ($\psi = \sqrt{r}$). Hence, from Eq. (19.47), since $p(r) = 0$ and $f(r) = \sqrt{r}$, the two linearly independent solutions are $\sqrt{r}$ and $\sqrt{r}\log r$.

Chapter 13

Problem 13.1 The WKB condition is

$$\int_0^a \sqrt{E_n - V(x)} \, dx = n\pi .$$

In this case we break the domain into three parts, and the integral becomes

$$\int_0^a \sqrt{E_n - V(x)} \, dx = \int_0^{\frac{a}{3}} \sqrt{E_n - V(x)} \, dx + \int_{\frac{a}{3}}^{\frac{2a}{3}} \sqrt{E_n - V(x)} \, dx$$
$$+ \int_{\frac{2a}{3}}^a \sqrt{E_n - V(x)} \, dx$$

---

[2]Page 248, *Fundamentals of Differential Equation*, by R. K. Nagle, E. B. Saff and A.D. Snider, 5th Ed., Addison-Wesley (2001).

$$n\pi = \frac{2a}{3}\sqrt{E_n} + \frac{a}{3}\sqrt{E_n - V_0}\ .$$

In the absence of the perturbation; i.e., for the infinite well, energy $E_n^0 = \frac{n^2\pi^2}{a^2}$. Hence $a = \frac{n\pi}{\sqrt{E_n^0}}$. Substituting this value of a into the above equation, we get

$$\frac{2}{3}\frac{n\pi}{\sqrt{E_n^0}}\sqrt{E_n} + \frac{n\pi}{\sqrt{3E_n^0}}\sqrt{E_n - V_0} = n\pi\ .$$

Simplifying above

$$E_n = 5E_n^0 \pm \frac{4}{3}\sqrt{9(E_n^0)^2 - 3E_n^0 V_0} - \frac{V_0}{3}\ .$$

$V_0 \to 0$, the root $5E_n^0 - \frac{4}{3}\sqrt{9(E_n^0)^2 - 3E_n^0 V_0} - \frac{V_0}{3}$ gives $E_n = E_n^0$.

Problem 13.2

The half-harmonic oscillator potential is

$$V(x) = \begin{cases} \infty & x < 0 \\ \frac{1}{4}\omega^2 x^2 & x > 0 \end{cases}\ .$$

For this case the WKB condition is

$$\int \sqrt{E_n - V(x)}\,dx = \int_0^{x_{cl}} \sqrt{E_n - \frac{1}{4}\omega^2 x^2}\,dx$$

$$= \left(n - \frac{1}{4}\right)\pi \quad n = 1, 2, 3, \ldots\ .$$

x_{cl} is the classical turning point: $E_n = \frac{1}{4}\omega^2 x_{cl}^2$. From the above equation we obtain

$$\frac{\omega}{2}\int_0^{x_{cl}} \sqrt{x_{cl}^2 - x^2}\,dx = \frac{\omega\pi}{8}x_{cl} = \frac{\omega\pi}{8}\cdot\frac{4E_n}{\omega^2} = \frac{E_n\pi}{2\omega} = \left(n - \frac{1}{4}\right)\pi\ ,$$

which gives $E_n = \left(2n - \frac{1}{2}\right)\omega$. This is the exact result.

Problem 13.3

The potential is given by

$$V = \frac{\omega^2 x^2}{4}, \quad x \in (-\infty, \infty)\ .$$

In this case, $E_n = \frac{\omega^2}{4}x_{cl1}^2 = \frac{\omega^2}{4}x_{cl2}^2$, or equivalently $x_{cl1}^2 = x_{cl2}^2 = \frac{4E_n}{\omega^2}$. WKB condition is

$$\int_{x_{cl1}}^{x_{cl2}} \sqrt{E_n - \frac{1}{4}\omega^2 x^2}\,dx = 2\int_0^{x_{cl2}} \sqrt{E_n - \frac{1}{4}\omega^2 x^2}\,dx = \left(n - \frac{1}{2}\right)\pi\ .$$

This integral was evaluated in Problem (13.2). Hence, $\frac{E_n \pi}{\omega} = \left(n - \frac{1}{2}\right)\pi$, or

$$E_n = \left(n - \frac{1}{2}\right)\omega \quad n = 1, 2, 3, \ldots$$

the exact solution for the harmonic oscillator.

Problem 13.4

Our objective is to derive Eq. (13.18):

$$\frac{\partial I(\alpha)}{\partial \alpha} = -\frac{\sin^{-1}\left(\sqrt{\frac{\alpha}{E}}\,\tan y\right)}{2\sqrt{\alpha}} + \frac{\sin^{-1}\left(\sqrt{\frac{E+\alpha}{E}}\,\tan y\right)}{2\sqrt{E+\alpha}}.$$

We start from

$$\frac{\partial I(\alpha)}{\partial \alpha} = -\frac{1}{2}\int_{-\tan^{-1}\sqrt{E}}^{\tan^{-1}\sqrt{E}}\left(\frac{\tan^2 y}{\sqrt{E - \alpha \cdot \tan^2 y}}\right)dy$$

$$= -\frac{1}{2}\int_{-\tan^{-1}\sqrt{E}}^{\tan^{-1}\sqrt{E}}\left(\frac{\sec^2 y}{\sqrt{E - \alpha \cdot \tan^2 y}}\right)dy$$

$$+ \frac{1}{2}\int_{-\tan^{-1}\sqrt{E}}^{\tan^{-1}\sqrt{E}}\left(\frac{dy}{\sqrt{E - \alpha \cdot \tan^2 y}}\right)$$

$$= -\frac{1}{2\sqrt{\alpha}}\int\left(\frac{d(\tan y)}{\sqrt{\frac{E}{\alpha} - \tan^2 y}}\right) + \frac{1}{2}\int\left(\frac{\cos y\, dy}{\sqrt{E - (E+\alpha)\cdot \sin^2 y}}\right)$$

$$= -\frac{1}{2\sqrt{\alpha}}\sin^{-1}\left(\sqrt{\frac{\alpha}{E}}\,\tan y\right) + \frac{1}{2\sqrt{E+\alpha}}\sin^{-1}\left(\sqrt{\frac{E+\alpha}{E}}\,\tan y\right).$$

Problem 13.5

The SWKB condition is $\int_{x_1}^{x_2}\sqrt{E_s - W^2} = n\pi, n = 1, 2, 3 \cdots$. E_s is the SWKB energy eigenvalue, and x_1, x_2 are the turning points, where $E_s - W^2 = 0$.

The superpotential for the harmonic oscillator is $W = \omega/2$; thus, the potential is $V_s = W^2 - W' = \omega^2 x^2/4 - \omega/2$. Note that this is shifted down from the standard V used in the WKB method: $V_s = V - \omega/2$.

Now the SWKB integral, $\int_{x_1}^{x_2} \sqrt{E_s - W^2}$ looks like the WKB integral, $\int_{x_1}^{x_2} \sqrt{E - V}$. This is the same integral as in Problem 13.4, hence each SWKB energy is given by $E_{s,n}/\omega$. Setting this equal to $n\pi$, $E_{s,n} = n\omega$, $n = 0, 1, 2, 3 \ldots$. Note that this yields a zero ground state energy, as is required for (unbroken) SUSY. Comparing this to the WKB result of Problem 13.4, $E_n = (n - 1/2)\omega n = 1, 2, 3 \ldots = (n + 1/2)\omega n = 0, 1, 2, 3 \ldots$, the SWKB result is, as expected, shifted down $1/2\omega$ from the WKB result.

Problem 13.6

The WKB condition for the Morse potential is given by

$$\int \left[E - A^2 \left(1 - e^{-z}\right)^2 + Ae^{-z} \right]^{\frac{1}{2}} dz = n\pi. \tag{19.48}$$

With a change of variable $e^{-z} = y$, the left hand side of Eq. (19.48) becomes

$$-\int \left[E - A^2 \left(1 - y\right)^2 + Ay \right]^{\frac{1}{2}} dy/y$$

$$= -\int \left[E - A^2 \left(1 - y\right)^2 - A(1 - y) + A \right]^{\frac{1}{2}} dy/y$$

$$= -\int \left[E + A + \frac{1}{4} - \left\{ A\left(1 - y\right) + \frac{1}{2} \right\}^2 \right]^{\frac{1}{2}} dy/y$$

$$= -A \int \frac{dy}{y} \left[\underbrace{\left(\frac{E + A + \frac{1}{4}}{A^2} \right)}_{\beta^2} - \left\{ \left(1 - y\right) + \frac{1}{2A} \right\}^2 \right]^{\frac{1}{2}}.$$

With two more changes of variable: $(1 - y) + \frac{1}{2A} = u$ and $y = (1 - u) + \frac{1}{2A}$, the integral becomes

$$A \int \frac{\sqrt{\beta^2 - u^2}}{(1 - y) + \frac{1}{2A}} du \ .$$

Now setting $u = \beta \sin \theta$, the above integral becomes

$$A \beta^2 \int_{-\pi/2}^{\pi/2} \left(\frac{\cos^2 \theta}{1 - \beta \sin \theta + \frac{1}{2A}} \right) d\theta \ .$$

An integration then leads to

$$\frac{1}{2A\beta^2}\left[2\sqrt{-4A^2\left(\beta^2-1\right)+4A+1}\,\tan^{-1}\left(\frac{2A\beta-(2A+1)\tan\left(\frac{\theta}{2}\right)}{\sqrt{-4A^2\left(\beta^2-1\right)+4A+1}}\right)\right]$$
$$+\left[\frac{-2A\beta\cos(\theta)+2A\theta+\theta}{2A\beta^2}\right].$$

$$(19.49)$$

Setting $-4A^2\left(\beta^2-1\right)+4A+1=4A^2(1-\alpha^2)$, where $\alpha=\frac{\sqrt{E}}{A}$, we get

$$\frac{4A\sqrt{1-\alpha^2}\,\tan^{-1}\left(\frac{2A\beta-(2A+1)\tan\left(\frac{\theta}{2}\right)}{2A\sqrt{1-\alpha^2}}\right)-2A\beta\cos(\theta)+2A\theta+\theta}{2A\beta^2}.\ (19.50)$$

Substituting the limits for θ; i.e., $\pm\pi/2$, the term containing $\tan^{-1}$ in Eq. (19.50) gives

$$\underbrace{\tan^{-1}\left(\frac{2A\beta-(2A+1)}{2A\sqrt{1-\alpha^2}}\right)}_{\theta_1}-\underbrace{\tan^{-1}\left(\frac{2A\beta+(2A+1)}{2A\sqrt{1-\alpha^2}}\right)}_{\theta_2}.$$

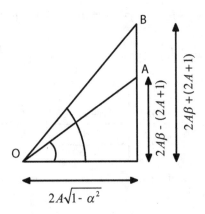

From the diagram above, the angle $\theta_1-\theta_2$ is the angle between vectors $\overrightarrow{OA}$ and $\overrightarrow{OB}$. Since

$$\overrightarrow{OA}=[2A\beta-(2A+1)]\hat{i}+[2A\sqrt{1-\alpha^2}]\hat{j}$$

and

$$\overrightarrow{OB}=[2A\beta+(2A+1)]\hat{i}+[2A\sqrt{1-\alpha^2}]\hat{j}\,,$$

we find that their scalar product is zero, hence the angle between them is $\theta_1 - \theta_2 = -\pi/2$. Thus, the Eq. (19.50), with limits put in, becomes

$$\frac{4A\sqrt{1-\alpha^2}(-\pi/2) + (2A+1)\pi}{2A\beta^2} . \qquad (19.51)$$

Substituting the above into our original integral of Eq. (19.49) and setting it equal to $\left(n + \frac{1}{2}\right)\pi$, we obtain

$$\frac{1}{2}\left(4A\sqrt{1-\alpha^2}(-\pi/2) + (2A+1)\pi\right) = \left(n + \frac{1}{2}\right)\pi .$$

Solving the above equation for energy, we get

$$E_n = A^2 - (A - n)^2 , \qquad (19.52)$$

the correct eigenvalues for the Morse potential.

Problem 13.7

The Morse superpotential $W(x, A, B) = A - Be^{-x}$, with a shift of variable $z = x - x_0$ with $x_0 = \log\left(\frac{B}{A}\right)$, can be written as $A\left(1 - e^{-z}\right)$. The SWKB integral for this superpotential can be written as

$$A \int \left[\frac{E}{A^2} - \left(1 - e^{-z}\right)^2\right]^{\frac{1}{2}} dz = A \int \left[\alpha^2 - \left(1 - e^{-z}\right)^2\right]^{\frac{1}{2}} dz$$

where $\alpha^2 \equiv E/A^2$. Let $e^{-z} = y$. Then $-e^{-z}dz = dy$ or $dz = -dy/y$.

$$A \int \left[\alpha^2 - \left(1 - e^{-z}\right)^2\right]^{\frac{1}{2}} dz = -A \int \left[\alpha^2 - (1 - y)^2\right]^{\frac{1}{2}} dy/y$$

$$= -A \int \frac{\sqrt{\alpha^2 - u^2}}{1 - u}(-du) , \qquad (19.53)$$

where we have used $1 - y = u$. Now setting $u = \alpha \sin\theta$, we get

$$I(\alpha) \equiv \int \frac{\sqrt{\alpha^2 - u^2}}{1 - u} du = \alpha^2 \int \frac{\cos^2\theta}{1 - \alpha\sin\theta} d\theta$$

$$= \frac{\sqrt{1-\alpha^2}(x - \alpha\cos(x)) - 2\left(\alpha^2 - 1\right)\tan^{-1}\left(\frac{\alpha - \tan\left(\frac{x}{2}\right)}{\sqrt{1-\alpha^2}}\right)}{\sqrt{1-\alpha^2}} .$$

Since the limits are $\pm\pi/2$, the integral $I(\alpha)$ is

$$
= \frac{\sqrt{1-\alpha^2}(\pi/2 - \alpha\cos(\pi/2)) - 2\left(\alpha^2 - 1\right)\tan^{-1}\left(\frac{\alpha - \tan\left(\frac{\pi}{4}\right)}{\sqrt{1-\alpha^2}}\right)}{\sqrt{1-\alpha^2}}
$$

$$
- \frac{\sqrt{1-\alpha^2}(-\pi/2 - \alpha\cos(-\pi/2)) - 2\left(\alpha^2 - 1\right)\tan^{-1}\left(\frac{\alpha - \tan\left(\frac{-\pi}{4}\right)}{\sqrt{1-\alpha^2}}\right)}{\sqrt{1-\alpha^2}}
$$

$$
= \frac{\sqrt{1-\alpha^2}(\pi/2) - 2\left(\alpha^2 - 1\right)\tan^{-1}\left(\frac{\alpha - 1}{\sqrt{1-\alpha^2}}\right)}{\sqrt{1-\alpha^2}}
$$

$$
- \frac{\sqrt{1-\alpha^2}(-\pi/2) - 2\left(\alpha^2 - 1\right)\tan^{-1}\left(\frac{\alpha + 1}{\sqrt{1-\alpha^2}}\right)}{\sqrt{1-\alpha^2}}
$$

$$
= \pi + \frac{-2\left(\alpha^2 - 1\right)\tan^{-1}\left(\frac{\alpha - 1}{\sqrt{1-\alpha^2}}\right)}{\sqrt{1-\alpha^2}} - \frac{-2\left(\alpha^2 - 1\right)\tan^{-1}\left(\frac{\alpha + 1}{\sqrt{1-\alpha^2}}\right)}{\sqrt{1-\alpha^2}}
$$

$$
= \pi - \frac{2\left(\alpha^2 - 1\right)}{\sqrt{1-\alpha^2}}\left(\tan^{-1}\left(\frac{\alpha - 1}{\sqrt{1-\alpha^2}}\right) - \tan^{-1}\left(\frac{\alpha + 1}{\sqrt{1-\alpha^2}}\right)\right)
$$

$$
= \pi + \frac{2\left(\alpha^2 - 1\right)}{\sqrt{1-\alpha^2}}\left(\tan^{-1}\left(\frac{-\alpha + 1}{\sqrt{1-\alpha^2}}\right) + \tan^{-1}\left(\frac{\alpha + 1}{\sqrt{1-\alpha^2}}\right)\right)
$$

$$
= \pi + \frac{2\left(\alpha^2 - 1\right)}{\sqrt{1-\alpha^2}}\underbrace{\left(\tan^{-1}\left(\frac{\sqrt{1-\alpha}}{\sqrt{1+\alpha}}\right) + \tan^{-1}\left(\frac{\sqrt{1+\alpha}}{\sqrt{1-\alpha}}\right)\right)}_{\pi/2}
$$

$$
= \pi + \frac{\left(\alpha^2 - 1\right)}{\sqrt{1-\alpha^2}}\pi = \pi\left(1 - \sqrt{1-\alpha^2}\right) .
$$

The SWKB condition then yields

$$
A\pi\left(1 - \sqrt{1-\alpha^2}\right) = n\pi . \tag{19.54}
$$

Replacing α^2 by E_n/A^2, we obtain the energy for the Morse potential:

$$
E_n = A^2 - (A - n)^2 .
$$

Chapter 14

Problem 14.1

From Eq. (14.38), we have

$$W^2(r, \alpha_0) + \frac{dW(r, \alpha_0)}{dr} = W^2(r, \alpha_1) - \frac{dW(r, \alpha_1)}{dr} + g(\alpha_1) - g(\alpha_0).$$

To see the singularity structure at infinity, we change variable to $u = \frac{1}{r}$.

$$\frac{dW(r, \alpha_0)}{dr} = \frac{dW(r(u), \alpha_0)}{du} \frac{du}{dr} = -u^2 \frac{dW(1/u, \alpha_0)}{du} \ .$$

This leads to the following shape invariance condition:

$$W^2\left(\frac{1}{u}, \alpha_0\right) - u^2 \frac{dW\left(\frac{1}{u}, \alpha_0\right)}{du} = W^2\left(\frac{1}{u}, \alpha_1\right) + u^2 \frac{dW\left(\frac{1}{u}, \alpha_1\right)}{du} + R(\alpha_0) \ ,$$

$$(19.55)$$

where $R(\alpha_0) = g(\alpha_1) - g(\alpha_0)$.

Problem 14.2

From Chapter 14, we obtain

$$W^2(1/u, \alpha_0) = \frac{(c_{-1}(\alpha_0))^2}{u^2} + \frac{(2 c_{-1}(\alpha_0) c_0(\alpha_0))}{u}$$
$$+ \left((c_0(\alpha_0))^2 + 2 c_{-1}(\alpha_0) c_1(\alpha_0)\right) + \cdots$$

and

$$-u^2 \frac{dW(1/u, \alpha_0)}{du} = c_{-1}(\alpha_0) - c_1(\alpha_0) u^2 - 2 c_2(\alpha_0) u^3 + \cdots \ .$$

Substituting these in Eq. (19.55) we obtain

$$\begin{bmatrix} c_{-1}(\alpha_0)^2 u^{-2} + \\ 2 c_{-1}(\alpha_0) c_0(\alpha_0) u^{-1} + \\ c_0(\alpha_0)^2 + 2 c_{-1}(\alpha_0) c_1(\alpha_0) \\ + c_{-1}(\alpha_0) - c_1(\alpha_0) u^2 + \cdots \end{bmatrix} - \begin{bmatrix} c_{-1}(\alpha_1)^2 u^{-2} + \\ 2 c_{-1}(\alpha_1) c_0(\alpha_1) u^{-1} + \\ c_0(\alpha_1)^2 + 2 c_{-1}(\alpha_1) c_1(\alpha_1) \\ - c_{-1}(\alpha_1) + c_1(\alpha_1) u^2 + \cdots \\ + g(\alpha_1) - g(\alpha_0) \end{bmatrix} = 0 \ .$$

Setting the coefficient of u^{-2} to zero, we obtain $c_{-1}(\alpha_0)^2 = c_{-1}(\alpha_1)^2$. The solution of this constraint is $c_{-1}(\alpha_0) = c_{-1}(\alpha_1)$. Then, the coefficient of u^{-1} gives $c_0(\alpha_0) = c_0(\alpha_1)$. Now from the u-independent terms we get

$$2 c_{-1}(\alpha_0) c_1(\alpha_0) + c_{-1}(\alpha_0) + g(\alpha_0) = 2 c_{-1}(\alpha_1) c_1(\alpha_1) - c_{-1}(\alpha_1) + g(\alpha_1) \ .$$

Problem 14.3

In this problem we were asked to derive the energy assuming that the origin was not a singular point. We assume the domain to be $(-\infty, \infty)$. This allows us to translate the origin by a finite amount without affecting the system. The expansion of W in powers of the coordinate r would then have the form

$$W(r, \alpha_0) = \sum_{k=1}^{\infty} b_k(\alpha_0) r^k = b_1(\alpha_0) r + b_2(\alpha_0) r^2 + \cdots .$$

The derivative of the superpotential is then given by

$$\frac{dW(r, \alpha_0)}{dr} = b_1(\alpha_0) + 2 b_2(\alpha_0) r + \cdots ;$$

and the square of the superpotential is given by

$$W^2(r, \alpha_0) = (b_1(\alpha_0))^2 r^2 + 2b_1(\alpha_0)) b_2(\alpha_0)) r^3 \cdots .$$

Substituting these expansions into the shape invariance condition (14.38), we arrive at the following constraints on coefficients of the superpotential $W(x, \alpha_0)$:

$$b_1(\alpha_0) + g(\alpha_0) = - b_1(\alpha_1) + g(\alpha_1)$$
$$b_2(\alpha_0) = -b_2(\alpha_1) \ . \tag{19.56}$$

These constraints are solved by

$$b_1(\alpha_0) = \omega \ , \text{ (a constant)}. \tag{19.57}$$

Then, $g(\alpha_0)$ is given by

$$g(\alpha_0) = \omega \, \alpha_0 \ , \tag{19.58}$$

and hence eigenvalues of the system are given by

$$E_n = g(\alpha_n) - g(\alpha_0) = g(\alpha_0 + n) - g(\alpha_0) = \omega \, (\alpha_0 + n) - \omega \, \alpha_0 \ = \ n\omega \ .$$

Problem 14.4

We want to show that the superpotential with singularity at the origin and no singularity at infinity is equivalent to the superpotential that is regular at the origin and has singularity at infinity. Here we consider the case that $b_{-1} = 0$ and $c_{-1} \neq 0$.

Near $r = 0$, the superpotential is given by

$$W(r, \alpha_0) = b_0(\alpha_0) + b_1(\alpha_0) r + \cdots . \tag{19.59}$$

From W, we obtain

$$\frac{dW}{dx} = \frac{dW}{dr}\frac{dr}{dx} = r\frac{dW}{dr} = b_1(\alpha_0)\,r + 2b_2(\alpha_0)\,r^2 + \cdots \;, \qquad (19.60)$$

and

$$W^2 = b_0(\alpha_0)^2 + 2b_0(\alpha_0)\,b_1(\alpha_0)\,r + \cdots \;. \qquad (19.61)$$

The shape invariance condition is

$$W^2(\alpha_0) + r\frac{dW(\alpha_0)}{dr} = W^2(\alpha_1) - r\frac{dW(\alpha_1)}{dr} + R(\alpha_0). \qquad (19.62)$$

Substituting the expansion of W into this equation and equating powers of r, we obtain

$$b_0^2(\alpha_0) + g(\alpha_0) = b_0^2(\alpha_1) + g(\alpha_1) \;. \qquad (19.63)$$

From Eq. (19.63), we get $b_0(\alpha_0) = \pm\sqrt{\gamma - g(\alpha_0)}$. We carry out a very similar analysis in the region $r \to \infty$, where the superpotential has the form

$$\widetilde{W}(u, \alpha_0) = \frac{c_{-1}(\alpha_0)}{u} + c_0(\alpha_0) + c_1(\alpha_0)\,u + c_1(\alpha_0)\,u^2 \cdots \;, \qquad (19.64)$$

and we obtain

$$c_{-1}(\alpha_0)^2 = c_{-1}(\alpha_1)^2 \Rightarrow c_{-1}(\alpha_0) \text{ is independent of } \alpha_0 \;, \qquad (19.65)$$

$$2\,c_{-1}(\alpha_0)\,c_0(\alpha_0) + c_{-1}(\alpha_0) = 2\,c_{-1}(\alpha_1)\,c_0(\alpha_1) - c_{-1}(\alpha_1)$$

$$\Rightarrow \quad c_0(\alpha_0) + 1 = c_0(\alpha_0 + 1)$$

$$\Rightarrow \quad c_0(\alpha_0) = \alpha_0\,. \qquad (19.66)$$

Thus, near $r = 0$

$$W(r, \alpha_0) = \pm\sqrt{\gamma - g(\alpha_0)} + b_1(\alpha_0)r + \cdots \qquad (19.67)$$

and near $r = \infty$,

$$\widetilde{W}(u, \alpha_0) = \frac{c_{-1}}{u} + \alpha_0 + \cdots \;. \qquad (19.68)$$

Near $r = 0$, the QHJ equation is given by

$$p^2 - i\,r\frac{dp}{dr} = E - W^2(r, \alpha_0) + r\frac{dW(r, \alpha_0)}{dr}, \qquad (19.69)$$

where the superpotential $W(r, \alpha_0)$ is given by Eq. (14.82). Expanding the quantum momentum function as

$$p(r) = \frac{p_{-1}}{r} + p_0 + p_1 r + \cdots , \qquad (19.70)$$

and substituting it in the above QHJ equation, we obtain

$$p_{-1}^2 = 0; \quad 2p_{-1}p_0 + i\,p_{-1} = 0; \quad p_0^2 = E - b_0(\alpha_0)^2 ; \quad \text{etc.} \quad (19.71)$$

Thus, we obtain $p_{-1} = 0$ and $p_0 = \pm\sqrt{E - b_0(\alpha_0)^2} = \pm\sqrt{E - (\gamma - g(\alpha_0))}$.

Near $r \to \infty$, we expand the QMF as $\tilde{p}(u) = \tilde{p}_{-1}u^{-1} + \tilde{p}_0 + \tilde{p}_1 u + \cdots$. We substitute this expansion for $\tilde{p}(u)$ and the superpotential given by Eq. (14.83) into the QHJ equation

$$\tilde{p}^2 + i\,u\frac{d\tilde{p}}{du} = E - \left(\widetilde{W}^2(u, \alpha_0) + u\frac{d\widetilde{W}(u, \alpha_0)}{du} \right) . \qquad (19.72)$$

Now comparing various powers of u, we obtain $\tilde{p}_{-1} = ic_{-1}$, $\tilde{p}_0 = i(\alpha_0 + 1)$, etc.

The quantization condition for this system then yields

$$\oint_{x=-\infty} p\, dx + \oint_{x=\infty} p\, dx = -2\pi n . \qquad (19.73)$$

Since near $x = -\infty$, the variable $r = e^x \to 0$ and as $x \to \infty$, the variable $u \equiv 1/r = e^{-x} \to 0$, we write the above condition as

$$\oint_{r=0} \left(\frac{p_0}{r} \right) dr - \oint_{u=0} \left(\frac{\tilde{p}_0}{u} \right) du = -2\pi n .$$

This then gives

$$\pm 2\pi i \sqrt{E_n - (\gamma - g(\alpha_0))} - 2\pi i \cdot i(\alpha_0 + 1) = -2\pi n .$$

Solving for E_n we obtain

$$E_n = (\gamma - g(\alpha_0)) - (1 + \alpha_0 + n)^2 .$$

Since we assume that supersymmetry is unbroken, we must have $E_0 = 0$. This implies $g(\alpha_0) = \gamma - (1 + \alpha_0)^2$, and hence $g(\alpha_n) = \gamma - (1 + \alpha_0 + n)^2$. Thus, we have

$$E_n = g(\alpha_n) - g(\alpha_0) = (1 + \alpha_0)^2 - (1 + \alpha_0 + n)^2 . \qquad (19.74)$$

If we identify $1 + \alpha_0$ with the parameter $-A$, we obtain the energy $E_n = A^2 - (A - n)^2$.

Chapter 15

Problem 15.1

Since $\sigma_x = \frac{S_x}{2}$, we want to show that $\chi_x^+ = \frac{1}{\sqrt{2}} \begin{pmatrix} 1 \\ 1 \end{pmatrix}$ and $\chi_x^- = \frac{1}{\sqrt{2}} \begin{pmatrix} 1 \\ -1 \end{pmatrix}$

are eigenvectors of $\sigma_x = \begin{pmatrix} 0 & 1 \\ 1 & 0 \end{pmatrix}$ with eigenvalues 1 and -1 respectively.

$$\sigma_x \chi_x^\pm = \begin{pmatrix} 0 & 1 \\ 1 & 0 \end{pmatrix} \frac{1}{\sqrt{2}} \begin{pmatrix} 1 \\ \pm 1 \end{pmatrix} = \pm \frac{1}{\sqrt{2}} \begin{pmatrix} 1 \\ \pm 1 \end{pmatrix} = \pm \chi_x^\pm \ .$$

Similarly, $\chi_y^+ = \frac{i}{\sqrt{2}} \begin{pmatrix} 1 \\ i \end{pmatrix}$ and $\chi_y^+ = \frac{i}{\sqrt{2}} \begin{pmatrix} i \\ 1 \end{pmatrix}$ are eigenvectors of $\sigma_y = \begin{pmatrix} 0 & -i \\ i & 0 \end{pmatrix}$ with eigenvalues ± 1.

Problem 15.2

We want to derive the anti-commutation relations:

$$\alpha_i \alpha_j + \alpha_j \alpha_i = 0, \quad j \neq i \quad \text{and} \quad \alpha_i \beta + \beta \alpha_i = 0, \quad i = 1, 2, 3 \,.$$

In Eq. (15.4) we obtained

$$p^2 + m^2 = \sum_{i=1}^{3} \alpha_i p_i \sum_{j=1}^{3} \alpha_j p_j + m^2 \beta^2 + m \left[\left(\sum_{i=1}^{3} \alpha_i p_i \right) \beta + \beta \left(\sum_{j=1}^{3} \alpha_j p_j \right) \right]$$

$$= \sum_{i=1}^{3} (\alpha_i p_i)^2 + \sum_{j \neq i} \alpha_i \alpha_j p_i p_j + m^2 \beta^2 + m \sum_{i=1}^{3} p_i (\alpha_i \beta + \beta \alpha_i) \ .$$

$$(19.75)$$

Since $p^2 + m^2$ only has diagonal terms of the form $\sum_i (p_i)^2$, the coefficients of the cross terms must vanish. Hence, $\sum_{j \neq i} \alpha_i \alpha_j p_i p_j = 0$. We need to be careful about this constraint to insure that we do not double count.

$$\sum_{j \neq i} \alpha_i \alpha_j p_i p_j = \left(\sum_{j < i} \alpha_i \alpha_j + \sum_{j > i} \alpha_i \alpha_j \right) p_i p_j \ .$$

$$= \left(\sum_{j < i} \alpha_i \alpha_j + \sum_{j < i} \alpha_j \alpha_i \right) p_i p_j$$

$$= \sum_{j < i} (\alpha_i \alpha_j + \alpha_j \alpha_i) p_i p_j \ , \quad (19.76)$$

where we have used dumminess of the indices i and j, and the symmetry of the factor $p_i p_j$ under the interchange of indices. For the last sum of Eq. (19.76) to be zero, we must have

$$\alpha_i \alpha_j + \alpha_j \alpha_i = 0.$$

Similarly, since $p^2 + m^2$ does not have a term linear in momentum, the last term of Eq. (19.75) must vanish as well. This implies

$$\alpha_i \beta + \beta \alpha_i = 0 .$$

Problem 15.3

We want to show that $[\vec{L}, H] = i\,(\vec{\alpha} \times \vec{p})$.

$$[L_x, H] = [L_x, \alpha_x p_x + \alpha_y p_y + \alpha_z p_z] + [L_x, \beta m] + \left[L_x, -\frac{e^2}{r}\right] .$$

The operator L_x contains coordinates and their derivatives, and hence commutes with β; i.e., $[L_x, \beta m] = 0$. The angular momentum operator L_x commutes with any spherically symmetric function of coordinates; i.e., $[L_x, V(r)] = 0$. So

$$[L_x, H] = \alpha_x \underbrace{[L_x, p_x]}_{0} + \alpha_y \underbrace{[L_x, p_y]}_{i\,p_z} + \alpha_z \underbrace{[L_x, p_z]}_{-i\,p_y}$$

$$= i\,(\alpha_y p_z + \alpha_z p_y) . \tag{19.77}$$

Generalizing to the other components of $\vec{L}$

$$\left[\vec{L}, H\right] = i\,(\vec{\alpha} \times \vec{p}) .$$

Problem 15.4

We want to show that

$$[\underline{\vec{\sigma}}, H] = [\underline{\vec{\sigma}}, \vec{\alpha} \cdot \vec{p} + \beta m + V(\vec{r})] = -2i(\vec{\alpha} \times \vec{p}) .$$

Since the operator $\underline{\sigma}_i \equiv \begin{pmatrix} \sigma_i & 0 \\ 0 & \sigma_i \end{pmatrix}$ commutes with the operator $\beta = \begin{pmatrix} 1 & 0 \\ 0 & -1 \end{pmatrix}$, and with the scalar potential $V(\vec{r})$, we have

$$[\underline{\vec{\sigma}}, H] = [\underline{\vec{\sigma}}, \vec{\alpha} \cdot \vec{p}\,] = [\underline{\vec{\sigma}}, \alpha_x p_x + \alpha_y p_y + \alpha_z p_z] .$$

Now

$$[\sigma_x, \alpha_x] = \begin{pmatrix} \sigma_x & 0 \\ 0 & \sigma_x \end{pmatrix} \begin{pmatrix} 0 & \sigma_x \\ \sigma_x & 0 \end{pmatrix} - \begin{pmatrix} 0 & \sigma_x \\ \sigma_x & 0 \end{pmatrix} \begin{pmatrix} \sigma_x & 0 \\ 0 & \sigma_x \end{pmatrix}$$

$$= \begin{pmatrix} 0 & \sigma_x^2 \\ \sigma_x^2 & 0 \end{pmatrix} - \begin{pmatrix} 0 & \sigma_x^2 \\ \sigma_x^2 & 0 \end{pmatrix} = 0 \ . \tag{19.78}$$

Similarly,

$$[\sigma_x, \alpha_y] = \begin{pmatrix} \sigma_x & 0 \\ 0 & \sigma_x \end{pmatrix} \begin{pmatrix} 0 & \sigma_y \\ \sigma_y & 0 \end{pmatrix} - \begin{pmatrix} 0 & \sigma_y \\ \sigma_y & 0 \end{pmatrix} \begin{pmatrix} \sigma_x & 0 \\ 0 & \sigma_x \end{pmatrix}$$

$$= \begin{pmatrix} 0 & \sigma_x\sigma_y \\ \sigma_x\sigma_y & 0 \end{pmatrix} - \begin{pmatrix} 0 & \sigma_y\sigma_x \\ \sigma_y\sigma_x & 0 \end{pmatrix}$$

$$= \begin{pmatrix} 0 & [\sigma_x, \sigma_y] \\ [\sigma_x, \sigma_y] & 0 \end{pmatrix} = i \begin{pmatrix} 0 & 2\sigma_z \\ 2\sigma_z & 0 \end{pmatrix} = 2i\,\alpha_z \ , \tag{19.79}$$

and

$$[\sigma_x, \alpha_z] = \begin{pmatrix} 0 & [\sigma_x, \sigma_z] \\ [\sigma_x, \sigma_z] & 0 \end{pmatrix} = -i \begin{pmatrix} 0 & 2\sigma_y \\ 2\sigma_y & 0 \end{pmatrix} = -2i\,\alpha_y \ .$$

Similarly, we obtain

$$[\sigma_y, \alpha_y] = 0, \quad [\sigma_y, \alpha_x] = -2i\,\sigma_z, \quad \text{and} \quad [\sigma_y, \alpha_z] = 2i\,\alpha_x,$$

and

$$[\sigma_z, \alpha_z] = 0, [\sigma_z, \alpha_x] = 2i\,\alpha_y, \quad \text{and} \quad [\sigma_z, \alpha_y] = -i\,\alpha_x \ .$$

Putting all of these together

$$[\sigma_x, H] = -2i\,(\vec{\alpha} \times \vec{p})_x \ .$$

Generalizing this result, we obtain

$$[\vec{\sigma}, H] = -2i\,(\vec{\alpha} \times \vec{p}) \ .$$

Problem 15.5

The exact fine structure is

$$E_{nj} = m \left\{ \left[1 + \left(\frac{\alpha}{n - (j + \frac{1}{2}) + \sqrt{(j + \frac{1}{2})^2 - \alpha^2}} \right)^2 \right]^{-\frac{1}{2}} \right\} . \tag{19.80}$$

Since $\alpha \ll 1$, $\sqrt{(j + 1/2)^2 - \alpha^2} = (j + 1/2)\sqrt{1 - \frac{\alpha^2}{(j+1/2)^2}} \approx (j + 1/2) - \frac{\alpha^2}{2(j+1/2)}$.

Canceling the $j + 1/2$'s in the denominator, and pulling n out, the term in big round brackets of Eq. 19.80 becomes $\frac{\alpha}{n}\left[1 + \frac{\alpha^2}{2n(j+1/2)}\right]$. Invoking again the smallness of α, the entire term in the curly brackets becomes, to order α^4, $1 - \frac{\alpha^2}{2n^2}\left[1 + \frac{\alpha^2}{n(j+1/2)}\right] + \frac{3\alpha^4}{8n^4}$. Collecting terms in powers of α, this is $1 - \frac{\alpha^2}{2n^2} + \frac{\alpha^4}{2n^4}\left(\frac{-n}{j+1/2} + \frac{3}{4}\right)$. This is now reduced to the Bohr fine structure:

$$\frac{-13.6eV}{n^2}\left[1 + \frac{\alpha^2}{n^2}\left(\frac{n}{j+1/2} - \frac{3}{4}\right)\right].$$

Subtracting the rest energy, the Bohr-model fine structure is

$$\frac{-13.6eV}{n^2}\left[\frac{\alpha^2}{n^2}\left(\frac{n}{j+1/2} - \frac{3}{4}\right)\right].$$

Problem 15.6

$$A^- A^+ \mathcal{F} = \left(-\frac{d}{d\mu} + \frac{s}{\mu} - \frac{\gamma}{s}\right)\left(\frac{d}{d\mu} + \frac{s}{\mu} - \frac{\gamma}{s}\right)\mathcal{F}$$

$$= -\frac{d^2\mathcal{F}}{d\mu^2} + \left(\frac{s}{\mu} - \frac{\gamma}{s}\right)^2 + \left[\left(\frac{s}{\mu} - \frac{\gamma}{s}\right) - \left(\frac{s}{\mu} - \frac{\gamma}{s}\right)\right]\frac{d\mathcal{F}}{d\mu} - \frac{s}{\mu^2}\mathcal{F}$$

$$= -\frac{d^2}{d\mu^2}\mathcal{F} + \left(\frac{s(s+1)}{\mu^2} - \frac{2\gamma^2}{\mu} + \frac{2s}{\mu^2}\right)\mathcal{F}.$$

Since $A^- A^+ \mathcal{F} = \left(\frac{\gamma^2}{s^2} + 1 - \frac{m^2}{E^2}\right)\mathcal{F}$

$$A^- A^+ \mathcal{F} - \left(\frac{\gamma^2}{s^2} + 1 - \frac{m^2}{E^2}\right)\mathcal{F} = -\frac{d^2}{d\mu^2}\mathcal{F} + \left(\frac{s(s+1)}{\mu^2} - \frac{2\gamma^2}{\mu} - 1 + \frac{m^2}{E^2}\right)\mathcal{F} = 0.$$

A similar calculations yields

$$-\frac{d^2\mathcal{G}}{d\mu^2} + \left(\frac{s(s+1)}{\mu^2} - \frac{2\gamma^2}{\mu} - 1 + \frac{m^2}{E^2}\right)\mathcal{G} = 0.$$

Chapter 16

Problem 16.1

From Eq. (16.3) we have

$$\left(\frac{d}{dx} + W(x)\right)\psi_E^-(x) = N_1\,\psi_E^+(x) \quad \text{and} \quad \left(-\frac{d}{dx} + W(x)\right)\psi_E^+(x) = N_2\,\psi_E^-(x),$$

where N_1 and N_2 are constants. Applying these conditions to the asymptotic forms given in Eq. (16.2), we obtain

$$\left(\frac{d}{dx} + W_{-\infty}\right)\left(e^{ikx} + r^-(k)e^{-ikx}\right) = N_1\left(e^{ikx} + r^+(k)e^{-ikx}\right),$$

and

$$\left(\frac{d}{dx} + W_\infty\right)t^- e^{ik'x} = N_1 t^+(k)e^{ik'x}.$$

These lead to

$$(ik + W_{-\infty})e^{ikx} + r^-(-ik + W_{-\infty})e^{-ikx} = N_1\left(e^{ikx} + r^+e^{-ikx}\right),$$
(19.81)

and

$$(ik' + W_\infty)t^-(k') = N_1 t^+(k').$$
(19.82)

Comparing the coefficients of e^{ikx} in Eq. (19.81), and using Eq. (19.82), we obtain following relationships:

$$(ik + W_{-\infty}) = N_1 \quad \text{and} \quad t^+(k') = \frac{(ik' + W_\infty)}{N_1}t^-(k').$$
(19.83)

Eliminating the coefficient N_1,

$$t^+(k') = \left(\frac{ik' + W_\infty}{ik + W_{-\infty}}\right)t^-(k').$$
(19.84)

Problem 16.2

From Eq. (16.5), we have

$$(-ik + W_{-\infty})e^{ikx} + r^+(ik + W_{-\infty})e^{-ikx} = N_2\left(e^{ikx} + r^- e^{-ikx}\right).$$
(19.85)

Comparing the coefficients of $e^{\pm ikx}$ in Eq. (16.4), we get following relationships:

$$(-ik + W_{-\infty}) = N_2 \quad \text{and} \quad r^+(ik + W_{-\infty}) = N_2 r^-.$$
(19.86)

Elimination of N_2 between these two equations leads to

$$r^+ = \left(\frac{W_{-\infty} - ik}{W_{-\infty} + ik}\right)r^-; \qquad r^- = \left(\frac{W_{-\infty} + ik}{W_{-\infty} - ik}\right)r^+.$$
(19.87)

These are same as those in Eq. (16.7).

Problem 16.3

(1) The partner potentials $V_\pm(x, a_0)$ generated by the superpotential $W(x, a_0) = a_0 \tanh x$ are:

$$\begin{aligned}
V_\pm(x, a_0) &= a_0^2 \tanh^2 x \pm a_0 \mathrm{sech}^2 x \\
&= a_0^2(1 - \mathrm{sech}^2 x) \pm a_0 \mathrm{sech}^2 x \\
&= a_0^2 - a_0(a_0 \mp 1)\, \mathrm{sech}^2 x)\ .
\end{aligned}$$

(2) Using Mathematica, we have plotted $W(x, 1)$ and $V_\pm(x, 1)$ for $-5 < x < 5$. The potential $V_+(x, 1)$ is equal to unity.

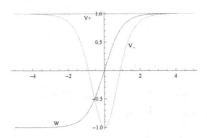

(3) For $a_0 = 1$, since the potential $V_+(x, a_0)$ is a constant, the coefficients $r^+(k, a_0)$ is zero. This implies that $r^-(k, a_0)$ is zero as well, and hence $t^-(k, a_0) = 1$. Thus, the reflection amplitude for $V_-(x, a_0)$ is zero for all values of energy, provided $a_0 = 1$.

Problem 16.4

Starting from

$$j_\ell(\rho) = (-1)^\ell \rho^\ell \left(\frac{1}{\rho}\frac{d}{d\rho}\right)^\ell \frac{\sin\rho}{\rho} \text{ and } n_\ell(\rho) = (-1)^{\ell+1}\rho^\ell \left(\frac{1}{\rho}\frac{d}{d\rho}\right)^\ell \frac{\cos\rho}{\rho}, \text{ we need}$$

to derive the leading behavior of the spherical Bessel functions near the origin.

We can write the above expression as

$$j_\ell(\rho) = (-1)^\ell \rho^\ell 2^\ell \left(\frac{d}{d\rho^2}\right)^\ell \sum_{k=0}^{\infty} \frac{(-1)^k \left(\rho^2\right)^k}{(2k+1)!}.$$

The leading term in the infinite sum that contributes after differentiation is $\frac{(-1)^\ell \rho^{2\ell}}{(2\ell+1)!}$. Since $\left(\frac{d}{d\rho^2}\right)^\ell \rho^{2k} = \frac{k!}{(k-\ell)!}\rho^{2(k-\ell)}$, setting $k = \ell$, we get $\left(\frac{d}{d\rho^2}\right)^\ell \rho^{2\ell} = \frac{\ell!}{0!} = \ell!$. Thus,

$$\lim_{\rho \to 0} j_\ell(\rho) \to = \frac{2^\ell \, \ell!}{(2\ell + 1)!} \rho^\ell.$$

Now let us consider the spherical Neumann function $n_l(\rho)$. We start with $n_\ell(\rho) = (-1)^{\ell+1} \rho^\ell \left(\frac{1}{\rho} \frac{d}{d\rho} \right)^\ell \frac{\cos \rho}{\rho}$. The leading term near $\rho = 0$ for $\left(\frac{1}{\rho} \frac{d}{d\rho} \right) \frac{\cos \rho}{\rho}$ is $\frac{(-1) \cos \rho}{\rho^3}$. Similarly, the leading term for $\left(\frac{1}{\rho} \frac{d}{d\rho} \right)^2 \frac{\cos \rho}{\rho}$ is $\frac{(-1)(-3) \cos \rho}{\rho^5}$, and so on. From this, we see that the leading term near $\rho = 0$ for $\left(\frac{1}{\rho} \frac{d}{d\rho} \right)^\ell \frac{\cos \rho}{\rho}$ is $\frac{(-1)(-3)(-5)\cdots(-2\ell+1) \, \cos \rho}{\rho^{2\ell+1}}$. This last expression can be written as $\frac{(-1)^\ell (2\ell)!}{2^\ell \ell!} \frac{1}{\rho^{2\ell+1}}$. Multiplying it by $(-1)^{\ell+1} \rho^\ell$, we get

$$\lim_{\rho \to 0} n_\ell(\rho) \to - \frac{(2\ell)!}{2^\ell \, \ell!} \frac{1}{\rho^{\ell+1}} \ .$$

Problem 16.5

For large values of ρ, the leading terms for $j_\ell(\rho)$ and $n_\ell(\rho)$ are obtained when all derivatives operate on the trigonometric functions. Using

$$\left(\frac{d}{d\rho} \right)^\ell \cos \rho = \cos \left(\rho + \frac{\ell \pi}{2} \right) = (-1)^\ell \cos \left(\rho - \frac{\ell \pi}{2} \right)$$

and

$$\left(\frac{d}{d\rho} \right)^\ell \sin \rho = \sin \left(\rho + \frac{\ell \pi}{2} \right) = (-1)^\ell \sin \left(\rho - \frac{\ell \pi}{2} \right)$$

we get the desired results.

Problem 16.6

The asymptotic form of the scattering states, far from the center of scattering, is

$$\frac{1}{kr} \left[A_\ell(k) \sin \left(kr - \frac{\ell \pi}{2} \right) - B_\ell(k) \cos \left(kr - \frac{\ell \pi}{2} \right) \right] \ . \tag{19.88}$$

We would like to write it in the form

$$\psi(r, \theta) = \frac{C_\ell(k)}{kr} \sin \left(kr - \frac{\ell \pi}{2} + \delta_\ell(k) \right) \ . \tag{19.89}$$

Expanding Eq. (19.89), we get

$$\psi(r, \theta) = \frac{C_\ell(k)}{kr} \left[\sin \left(kr - \frac{\ell \pi}{2} \right) \cos \delta_\ell(k) + \cos \left(kr - \frac{\ell \pi}{2} \right) \sin \delta_\ell(k) \right] \ . \tag{19.90}$$

The right hand sides of Eqs. (19.88) and (19.90) must be equal for all values of r. This is possible only if

$$A_\ell(k) = C_\ell(k) \cdot \cos \delta_\ell(k) \quad \text{and} \quad -B_\ell(k) = C_\ell(k) \cdot \sin \delta_\ell(k) \quad .$$

From these conditions, we get

$$\tan \delta_\ell(k) = -\left(\frac{B_\ell(k)}{A_\ell(k)}\right) \quad \text{and} \quad C_\ell(k) = \sqrt{A_\ell^2(k) + B_\ell^2(k)} \, .$$

Problem 16.7

From chapter 16, we have

$$\left[\frac{d}{dr} + W_\infty\right] \sin\left(kr - \frac{\ell\pi}{2} + \delta_\ell^{(-)}(k)\right) = N \sin\left(kr - \frac{\ell\pi}{2} + \delta_\ell^{(+)}(k)\right) \, .$$
$$(19.91)$$

Replacing $\sin\left(kr - \frac{\ell\pi}{2} + \delta_\ell^{(-)}(k)\right)$ by

$$\frac{1}{2i}\left(e^{i(kr - \frac{\ell\pi}{2} + \delta_\ell^{(-)}(k))} - e^{-i(kr - \frac{\ell\pi}{2} + \delta_\ell^{(-)}(k))}\right) ,$$

we get

$$\left[(ik + W_\infty) e^{i\delta_\ell^{(-)}(k)} - N e^{i\delta_\ell^{(+)}(k)}\right] e^{i\left(kr - \frac{\ell\pi}{2}\right)}$$
$$= \left[(-ik + W_\infty) e^{-i\delta_\ell^{(-)}(k)} + N e^{-i\delta_\ell^{(+)}(k)}\right] e^{-i\left(kr - \frac{\ell\pi}{2}\right)}.$$

This equation must hold for all values of r. Since $e^{\pm ikr}$ are linearly independent, we set their coefficients to zero. This yields following two conditions:

$$(ik + W_\infty) e^{i\delta_\ell^{(-)}(k)} = N e^{i\delta_\ell^{(+)}(k)} \text{ and } (-ik + W_\infty) e^{-i\delta_\ell^{(-)}(k)} = -N e^{-i\delta_\ell^{(+)}(k)}.$$

Dividing the second condition by the first, we get

$$e^{2i\delta_\ell^{(+)}} = \left[\frac{ik + W_\infty}{ik - W_\infty}\right] e^{2i\delta_\ell^{(-)}} \, . \tag{19.92}$$

Problem 16.8

Since $W = A \tanh r - B \coth r$, the asymptotic value of the superpotential $W_\infty = A - B$. Substituting this into Eq. (19.92), we obtain

$$e^{2i\delta_{B+1}^{(-)}} = -\left[\frac{A - B + ik}{A - B - ik}\right] e^{2i\delta_B^{(-)}} \, . \tag{19.93}$$

Since under reparametrization $B \to B + 1$, iterating the above equation B-times (if B were not an integer then by the maximum integer contained in it,) we obtain

$$e^{2i\delta_B^{(-)}} = (-)^B \frac{\Gamma(A - B + ik)}{\Gamma(A - B - ik)} \, \alpha_k \; .$$

Problem 16.9

Since $W = -A \coth r + \frac{B}{A}$, the asymptotic value of the superpotential, $W_\infty = \frac{B}{A} - A$. Substituting this into Eq. (19.92), we obtain

$$
\begin{aligned}
e^{2i\delta_{A+1}^{(-)}} &= - \left[\frac{\frac{B}{A} - A + ik}{\frac{B}{A} - A - ik} \right] e^{2i\delta_A^{(-)}} \\
&= - \frac{B - A^2 + ik\,A}{B - A^2 - ik\,A} \, e^{2i\delta_A^{(-)}} \\
&= - \frac{\left(A - \frac{ik}{2} \right)^2 - \left(B^2 + \frac{k^2}{4} \right)}{\left(A + \frac{ik}{2} \right)^2 - \left(B^2 + \frac{k^2}{4} \right)} \, e^{2i\delta_A^{(-)}} \\
&= - \frac{\left(A - \frac{ik}{2} + \sqrt{B^2 + \frac{k^2}{4}} \right) \left(A - \frac{ik}{2} - \sqrt{B^2 + \frac{k^2}{4}} \right)}{\left(A + \frac{ik}{2} + \sqrt{B^2 + \frac{k^2}{4}} \right) \left(A + \frac{ik}{2} - \sqrt{B^2 + \frac{k^2}{4}} \right)} \, e^{2i\delta_A^{(-)}} \; .
\end{aligned}
$$

$$(19.94)$$

Since under reparametrization $A \to A - 1$, iterating the above equation A-times (if A is not an integer then by the maximum integer contained in it), we get

$$e^{2i\delta_A^{(-)}} = (-)^A \frac{\Gamma\left(A - \frac{ik}{2} + \sqrt{B^2 + \frac{k^2}{4}} \right) \Gamma\left(A - \frac{ik}{2} - \sqrt{B^2 + \frac{k^2}{4}} \right)}{\Gamma\left(A + \frac{ik}{2} + \sqrt{B^2 + \frac{k^2}{4}} \right) \Gamma\left(A + \frac{ik}{2} - \sqrt{B^2 + \frac{k^2}{4}} \right)} \, \alpha_k \; .$$

Chapter 17

Problem 17.1

The constraint on values of α and β comes from Eq. (17.5).

$$z^{1+\beta}(1 - z)^{-\alpha-\beta} = \frac{2z(1 - z)}{\sqrt{R(z)}} \; ,$$

where $R(z) = az^2 + bz + c$. Solving for $R(z)$, we obtain

$$R(z) = 4z^2(1 - z)^2 z^{-2(1+\beta)}(1 - z)^{2(\alpha+\beta)} = 4z^{-2\beta}(1 - z)^{2(1+\alpha+\beta)} .$$

We must have

$$az^2 + bz + c = 4z^{-2\beta}(1 - z)^{2(1+\alpha+\beta)} .$$

First of all, since $R(z)$ is a quadratic function of z, β cannot be positive. The only allowed values for β are 0, $-\frac{1}{2}$ and -1. We will consider these two cases separately.

(1) $\beta = 0$. In this case, there are three possibilities for the combination $2(1 + \alpha + \beta) = 2(1 + \alpha)$. It can be 0, 1 or 2. Corresponding values for α are -1, $-\frac{1}{2}$, 0.

(2) $\beta = -\frac{1}{2}$. In this case, there are two possibilities for the combination $2(1 + \alpha + \beta) = 2(\frac{1}{2} + \alpha)$. It can be 0 or 1. Corresponding values for α are $-\frac{1}{2}$, 0.

(3) $\beta = -1$. In this case, there is only one allowed value for the combination $2(1 + \alpha + \beta) = 2(\alpha)$. It has to be equal to zero. Hence, $\alpha = 0$.

Summary of Possibilities
$\alpha = 0, \ \beta = 0, \ -\frac{1}{2}, \ -1$; $\quad \alpha = -\frac{1}{2}, \ \beta = 0, \ -\frac{1}{2}$; $\quad \alpha = -1, \ \beta = 0$.

Problem 17.2

Comparing this case of $\alpha = 0$, $\beta = -1$ with the case $\alpha = \beta = 0$ described in the text, we see that z and $1 - z$ simply exchange roles. Define $q \equiv 1 - z$ and the identical calculation occurs, producing the Eckart potential.

Problem 17.3

For $\alpha = -1$, $\beta = 0$. Eq. (17.5) becomes,

$$z(1 - z) = \frac{2z(1 - z)}{\sqrt{R(z)}},$$

whence $R(z) = 4$. The Natanzon potential is then

$$U[z(r)] = \frac{-fz(1 - z) + h_0(1 - z) + h_1 z}{4} + \frac{1}{4} . \qquad (19.95)$$

Defining $A \equiv \frac{-f}{4}, B \equiv \frac{h_0}{4}, C \equiv \frac{h_1}{4}$, with $z = [1 + \tanh\left(\frac{r}{2}\right)]/2$,

$$U = A \left(\frac{1 + \tanh\left(\frac{r}{2}\right)}{2} \right) \left(\frac{1 - \tanh\left(\frac{r}{2}\right)}{2} \right) + B \left(\frac{1 - \tanh\left(\frac{r}{2}\right)}{2} \right)$$

$$+ C \left(\frac{1 + \tanh\left(\frac{r}{2}\right)}{2} \right) + \frac{1}{4} \, .$$

Expanding the first term and collecting terms,

$$U = \frac{A}{4} - \frac{A}{4} \tanh^2 \left(\frac{r}{2} \right) + \frac{B + C}{2} + \frac{C - B}{2} \tanh \left(\frac{r}{2} \right) + \frac{1}{4}$$

$$= \frac{A + 1}{4} + \frac{B + C}{2} + \frac{1}{4} - \frac{A}{4} \tanh^2 \left(\frac{r}{2} \right) + \frac{C - B}{2} \tanh \left(\frac{r}{2} \right).$$

Since this must have the form $W^2 - \frac{dW}{dr}$, we examine various possibilities, and eventually choose

$$W = \tilde{m}_1 \tanh \left(\frac{r}{2} \right) + \tilde{m}_2.$$

This gives U, with the identifications

$$\tilde{m}_2^2 - \tilde{m}_2 = \frac{A + 2B + 2C + 1}{4}, \quad \tilde{m}_1^2 + 2\tilde{m}_2 = -\frac{A}{4}, \quad 2\tilde{m}_1 \tilde{m}_2 + \tilde{m}_2^2 = \frac{C - B}{2}.$$

This is the Rosen-Morse potential.

Further Readings

[1] R. Adhikari, R. Dutt, A. Khare, and U. Sukhatme, "Higher order WKB approximations in supersymmetric quantum mechanics", *Phys. Rev.* **A 38**, 1679 (1986).

[2] G. B. Arfken, H. J. Weber, F. Harris *Mathematical Methods for Physicists*, 5th edition, San Diego: Harcourt-Academic Press (2001).

[3] A. B. Balantekin, "Accidental degeneracies and supersymmetric quantum mechanics", *Ann. of Phys.* **164**, 277 (1985).

[4] A. B. Balantekin, "Algebraic approach to shape invariance", *Phys. Rev. A* **57**, 4188 (1998).

[5] H. Bethe and E. E. Salpeter, *Quantum Mechanics of One- and Two- Electron Atoms.* NY: Springer (1977).

[6] F. Cooper and B. Freedman, "Aspects of supersymmetric quantum mechanics", *Ann. Phys.* **146**, 262 (1983).

[7] F. Cooper, A. Khare and U. Sukhatme, "Supersymmetry and quantum mechanics", *Phys. Rep.* **251**, 268 (1995).

[8] F. Cooper, A. Khare, and U. Sukhatme, *Supersymmetry in Quantum Mechanics*, New Jersey: World Scientific (2001).

[9] P. A. M. Dirac, Communications of the Dublin Institute of Advanced Study, **A1**, 5-7 (1943).

[10] P. A. M. Dirac, *The Principles of Quantum Mechanics*, 4th edition, Oxford University Press (1958).

[11] R. Dutt, A. Khare, and U. P. Sukhatme: "Supersymmetry, shape invariance and exactly eolvable potentials", *Am. Jour. Phys.* **56**, 163 (1988).

[12] R. Dutt, A. Gangopadhyaya, and U. Sukhatme, "Noncentral potentials and spherical harmonics using supersymmetry and shape invariance", *Am. Jour. Phys.* **65**, 400 (1997).

[13] R. Dutt, A. Gangopadhyaya, C. Rasinariu, and U.P. Sukhatme; "New solvable singular potentials", *Jour. Phys. A-Math. & Gen.* **34**, 4129, (2001).

[14] V. A. Fock, *Zeits. f. Phys.* **49**, 339 (1928).

[15] W. M. Frank, D. J. Land and R. M. Spector, "Singular potentials", *Rev. Mod. Phys.* **43**, 36 (1971).

[16] L. E. Gendenshtein, "Derivation of exact spectra of the Schrödinger equation

273

       by means of supersymmetry", *Pismah. Eksp. Teor. Fiz.* **38**, 299 (1983); [English: JETP Lett. **38**, 356 (1983)].

[17] L. E. Gendenshtein and I. V. Krive: "Supersymmetry in quantum mechanics", *Usp. Fiz. Nauk* **146**, 553 (1985); [English: Sov. Phys. Usp. **28**, 645 (1985)].

[18] A. Gangopadhyaya, P. K. Panigrahi and U. P. Sukhatme, "Analysis of inverse-square potentials using supersymmetric quantum mechanics", *Jour. Phys. A-Math. & Gen.* **27**, 4295 (1994).

[19] A. Gangopadhyaya and U. P. Sukhatme; "Potentials with two shifted sets of equally spaced eigenvalues and their Calogero spectrum", *Phys. Lett. A* **224**, 5 (1996).

[20] A. Gangopadhyaya, J. Mallow, and U. Sukhatme, "Translational shape invariance and the inherent potential algebra", *Phys. Rev.* **A 58**, 4287 (1998).

[21] A. Gangopadhyaya, J. V. Mallow, and U. P. Sukhatme, "Broken supersymmetric shape invariant systems and their potential algebras", *Phys. Lett.* **A 283**, 279 (2001).

[22] A. Gangopadhyaya and J.V. Mallow, "Generating shape invariant potentials", *Int. Jour. of Mod. Phys.* **A 23**, 4959 (2008).

[23] Ya. I. Granovskii, "Sommerfeld formula and Dirac's theory", *UFN* **174**, 577 (2004).

[24] D. J. Griffiths, Quantum Mechanics, 2nd ed., CA: Benjamin-Cummings (2004).

[25] E. Gozzi, "Classical and quantum adiabatic invariants", *Phys. Lett.* **B 165**, 351 (1985).

[26] L. Infeld and T.E. Hull "The factorization method", *Rev. Mod. Phys.* **23**, 21 (1951).

[27] I. A. Ivanov, "WKB quantization of the Morse hamiltonian and periodic meromorphic functions", *Jour. Phys. A-Math. & Gen.* **30**, 3997 (1997).

[28] A. Jevicki and J. Rodrigues, "Singular potentials and supersymmetry breaking", *Phys. Lett.* **B 146**, 55 (1984).

[29] G. Junker, *Supersymmetric Methods in Quantum and Statistical Physics*, Berlin: Springer Verlag (1996).

[30] A. Khare, U. P. Sukhatme, "Scattering amplitudes for supersymmetric shape-invariant potentials by operator methods", *Jour. Phys. A-Math. & Gen.* **21**, L501 (1988).

[31] A. Khare and U. Sukhatme, "Phase equivalent potentials obtained from supersymmetry", *Jour. Phys. A-Math. & Gen.* **22**, 2847 (1989).

[32] L. D. Landau and E. M. Lifshitz, *Quantum Mechanics*, NY: Pergamon Press (1977).

[33] R. E. Langer, "On the connection formulas and solutions of the wave equation", *Phys. Rev.* **51**, 669 (1937).

[34] R. A. Leacock and M. J. Padgett, "Hamilton-Jacobi/action-angle quantum mechanics", *Phys. Rev.* **D 28**, 2491(1983).

[35] A. Messiah, *Quantum Mechanics*, v.1, Paris: Dunod Publ. (1964).

[36] G. A. Natanzon, "Study of the one-dimensional Schroedinger equation generated from the hypergeometric equation", Leningrad University Publication

[Vestnik Leningradskogo Universiteta], **10**, 22 (1971.); [English Translation by Haret Rosu is available at arXiv:physics/9907032v1].

[37] C. Rasinariu, J. V. Mallow and A. Gangopadhyaya, "Exactly solvable problems of quantum mechanics and their spectrum generating algebras: A review", *Central Euro. Jour. of Phys.* **5**, 111 (2007).

[38] H. Rosu and J. R. Guzman, "Gegenbauer polynomials and supersymmetric quantum mechanics", *Nuovo Cimento della Societa Italiana di Fisica* **B 112**, 941 (1997).

[39] P. Roy and R. Roychoudhury, "Question of degenerate states in supersymmetric quantum mechanics", *Phys. Rev.* **D 32**, 1597 (1985).

[40] E. Schrödinger, "Further studies on solving eigenvalue problems", *Proc. Roy, Irish Acad.* **46A**, 183-206(1941).

[41] U. Sukhatme, C. Rasinariu, A. Khare, "Cyclic shape invariant potentials", *Phys. Lett.* **A 234**, 401 (1997).

[42] A. Sukumar, "Supersymmetry and the Dirac equation for a central Coulomb field", *Jour. Phys. A-Math. & Gen.* **18**, L697 (1985).

[43] E. Witten, "Dynamical breaking of supersymmetry", *Nucl. Phys.* **B 188**, 513 (1981).

[44] E. Witten, "Constraints on supersymmetry breaking", *Nucl. Phys.* **B 202**, 253 (1982).

[45] J. Wu, Y. Alhassid, and F. Gürsey: "Group theory approach to scattering. IV. Solvable potentials associated with SO(2,2)", *Ann. Phys.* **196**, 163 (1989).

Index

Addendum

A new class of shape invariant potentials has been discovered [1, 2]. These potentials depend intrinsically on $\hbar$ and are not generated by the method presented in Chapter 11. There is a generalization of this method [3] which reproduces these and predicts others.

References

[1] C. Quesne, "Exceptional orthogonal polynomials, exactly solvable potentials and supersymmetry", *J. Phys.* **A 41**, 392001 (2008).

[2] S. Odake and R. Sasaki, "Another set of infinitely many exceptional (X_ℓ) Laguerre polynomials", *Phys. Lett.* **B 679**, 414 (2009).

[3] J. Bougie, A. Gangopadhyaya, and J. V. Mallow, "Generation of a complete set of supersymmetric shape invariant potentials from an Euler equation", http://arxiv.org/abs/1008.2035.